EARLY HISTORY

OREGON SEED PRODUCTION

Oregon State University Professor Emeritus

Donald H. Brewer

OSU SEED CERTIFICATION SERVICE

CORVALLIS, OREGON

Published by:
OSU Seed Certification Service
Corvallis, OR 97331
osu-cert@OSCS.ORST.EDU

CIP information available upon request

ISBN: 1-931979-08-1

Design and production by Kassell Concepts
Silverton, Oregon

Printed in USA

Contents

Introduction

Oregon's climatic conditions have contributed immensely to the seed quality that Oregon grown seed enjoys. Very few locations in the world are able to harvest directly out of the field and place straight into storage. In some cases the seed is harvested, conditioned, and shipped immediately. Most seed producing regions must harvest at a high rate of moisture, and artificially dry before warehousing. The artificial process renders a lower germination and is prone to attack by fungus, making the seed appear dark. Many reports tell of Oregon grown seed being blended with local grown seed in the U.S. or overseas to bring the quality to a higher standard.

The individuals mentioned in this book recognized early in the history of Oregon's seed production that our products could not be surpassed for quality. It is to their credit, too, that the state was acknowledged as having professional seed growers with the knowledge and ability to capitalize on the climatic conditions.

Dedication

This Book must be dedicated to my family. As with most Agriculture pursuits, the summer months are the busiest times, thus Dad was not around much. Occasionally it was possible to take the family to Ontario, Oregon and stay in a motel while dad made field inspections during the week. Sometimes two or more families would be assigned to the area, and that would be a vacation for the families. I must make special mention of my wife, Helen. She was able to keep the family together while I was away for extended periods of time; and later help with the entertaining of out-of-state and out-of-country guests.

I also dedicate this book to Ed Hardin. In fact, the entire Oregon Seed Program owes a great deal to Mr. Hardin. Ed's foremost task was running the State Seed Laboratory. He taught classes at Oregon State University and felt strongly that students could help make field inspections for seed certification in the summer, and be useful in the Seed Lab the rest of the year.

Through Ed Hardin's efforts, the Oregon State Seed Laboratory became world renowned. New innovations in seed testing techniques, equipment, and the accuracy of the test reports all were the result of Ed's leadership. No small feat when one considers the vast number of seed lots tested in a year's time.

I am indebted to those who have felt that this historical information is of merit. Thank you.

Don Brewer

Chapter I

Early Willamette Valley

Mr. E. R. Jackman, early Oregon State College (O.S.C.) Extension Agronomist, wrote the following concerning native grasses in Western Oregon: "Let's consider what the Willamette Valley was like before the white man came. The streams were pretty well choked up with beaver dams and other obstructions so that in the wintertime the water tended to spread out. In the earliest years of settlement, therefore, there were no roads through the center of the valley. The roads tended to follow both sides. There was one road from about the Molalla country on the south through Silverton and Sweet Home and so on down to Eugene. There was another road on the west side of the valley reaching from west of Hillsboro on down through where Independence is now and on to Philomath and south from there. This was because the center of the valley in many places was impassable in the winter time and, therefore, there was not much point in building roads there.

Traffic for a number of years was mostly by the rivers. There was a whole series of boat landings all the way through Clackamas county and boats even went up the Yamhill River for many years before Yamhill ever had a road across the hills to where Portland is.

The reason this is mentioned is that a large number of native grasses were of the moisture-loving types such as the Spike Red Top and Reed Canarygrass, and others. There were natural open prairies here and there, some of them on the wet ground such as around French Prairie and some of the prairies in Clackamas county, and some such as the Bald Hills in Yamhill county. Yamhill comes from a tribe of

Indians called *Yamel* and the word means *Bald Hills*. These prairies were kept in grass partly because the Indians burned some of them to make the hunting better and to facilitate their movement, and the prairies were partly due to the fact that they were wet in the wintertime so that most trees did not grow under those conditions."

Very few of the species grown for seed were native to Oregon. Most of the pasture species were introduced. For example, orchardgrass was found growing commonly throughout Oregon, but was known to be introduced from Europe. Orchardgrass was found growing in and around the ballast pits at Linnton, Oregon in 1916, however, the first certified seed was harvested in 1952. On the other hand, creeping bent-grass was identified growing at the same ballast site in 1916 and ten years later pure stands were harvested and certified.

The point being that seed production in Oregon did not begin all at once. It has been a slow progression up to the present time. All of this was helped along through the efforts of USDA investigators and Oregon State University Research and Extension personnel. Private seed interests have always encouraged seed production in Oregon.

Chapter II

Seed Certification in Oregon

Professor George Hyslop, member of the Farm Crops Department, Oregon Agriculture College, first undertook seed certification with potatoes in 1916 to provide seed stock true to variety name and reasonably free of diseases. Two years later, 1918, Oregon State Agricultural College started the certification of wheat. Although certification started in Oregon in 1916, the program was not legally vested in any organization until 1937 when a state law was passed officially giving Oregon State Agricultural College responsibility for the program. Specifically the Dean of Agriculture was given the responsibility and, of course, he was to hire assistants. The program was to be self sufficient and not cost the State taxpayers any money. In fact, the law specified that the money collected for services was to be dispensed by the Extension Service. Thus, the Seed Certification Project was placed under the direction of an extension specialist of the Federal Cooperative Extension Service and an advisory certification board.

The question of why seed certification was lodged with the College as an Extension Project has been asked many times. Nationally, in 1919, other states formed the International Crop Improvement Association and were well on their way to certifying crops before Oregon entered the national picture. Most of the state certification programs were under the auspices of the Land Grant Institution but were formed as Crop Improvement Associations governed by local growers. Some states on the other hand simply lodged their certification programs with the State Departments of Agriculture, and were run in a regulatory fashion. The key to answering the question of "why the Extension Service" lies with George Hyslop. His philosophy was that the certification program allowed for a great deal of grower contact; contacts that might not otherwise have been made. As we go along in this discussion it will be seen that George Hyslop boasted of many friends made while certifying bentgrass on the coast. Later, E.R. Jackman traveled the state making field inspections of potato and grain fields. In these men's minds it just followed that the seed certification work they were doing was truly Extension type work. That philosophy seems to have been carried on through to the present time.

Chapter III

Early Grass Seed Production

RYEGRASS

Floyd Mullen, one of Linn County's first County Extension Agents, in his book The Land of Linn states that "the time of the arrival of both annual and perennial ryegrass is somewhat debatable. Some reports indicated that ryegrass, as high as backs of cattle, was growing through the valley when the immigrants arrived. Indians had recognized ryegrass by naming a town in honor of this lush growing crop—Waiilatpu, Washington, location of the Whitman Mission. In Nez Perce language the word *Waiilatpu* means 'place of the big ryegrass.' Other reports, which can be verified, indicate the crop was introduced shortly before 1900."

Darnel ryegrass (*Lolium temulentum L*) can still be found in some of the uncultivated areas of the Willamette Valley and was probably the grass that was so tall reaching the backs of cattle. The ryegrass found in the drier regions of eastern Oregon and Washington was Giant Wild rye (*Elymus*).

DOMESTIC RYEGRASS

Domestic ryegrass, or common ryegrass, as grown today in the Willamette Valley, arrived before the turn of the century. It originated in Southern Europe and brought with it the original name of Italian ryegrass. As early as 1890, a Mr. John, an Englishman associated with the old Portland Seed Company, introduced into the northern Willamette Valley, a new strain of ryegrass which he had shipped in from Europe. The next year, 1891, William Felzer of Tangent obtained a small amount of seed from M. Senders and Co., Albany, who in turn had received it from the old Portland Seed Company. This may have been the actual beginning of the present day ryegrass, according to Mr. Mullen.

As early as 1910, grain farmers found domestic ryegrass to be a nuisance in grain crops because of its ability to make a volunteer growth without being planted. It still volunteers to this day. There were no markets for seed, and much of it was burned after being cleaned from grain. By 1915, the value of ryegrass as a pasture crop in other areas of the state was becoming recognized. The demand for seed for pasture purposes began to develop, particularly in the dairy areas of Tillamook County.

According to Mr. Mullen, the Jenks family of Tangent was one of the first to grow ryegrass seed on a commercial basis in Linn County. Forest Jenks perhaps was the first commercial seed producer. In the fall of 1921 he planted 40 acres solely to ryegrass which, were harvested for seed in 1922. The seed was cleaned by W.A. Vollstedt, who operated a seed cleaning plant in Tangent. The seed was purchased by Howard Jenks, Sr. of the Jenks-White Seed Company.

Howard Jenks, Sr., a partner in the pioneer J.E. Jenks and Son firm, was the first commercial shipper of domestic ryegrass seed. This firm operated a seed buying business and was instrumental in introducing several Oregon-grown seed crops to buyers and users in the East. To this firm goes the credit of opening an eastern market for Linn County's greatest crop - ryegrass seed.

Once ryegrass became recognized, several other farmers were quick to grab the opportunity of producing the new crop commercially. Chester Curtis and Mr. Langdon of Halsey produced seed in 1922, mainly from volunteer fields. By 1923 commercial production was well on its way. Thomas Jackson and Jess Skirvin of Harriburg, C.H. Davidson of Halsey, Harvey Pugh of Shedd, and Lee McLagan of Tangent were pioneer ryegrass seed growers in their areas.

PERENNIAL RYEGRASS

Perennial ryegrass, more commonly known as English ryegrass, with the name adopted from the country in which it originated, is just about as old as domestic ryegrass, but was an orphan for at least forty years. It is speculated that perennial ryegrass was one of those plants that established a foothold escaping from the ballast pits along the Columbia and Willamette Rivers. It has been reported here since 1890. It was mixed with domestic ryegrass and actually had no identity of its own.

Mr. Mullen states that not until 1930 did anyone seriously consider growing perennial ryegrass as a commercial crop. In the fall of that year C. P. Kizer of Harrisburg obtained from England, a small quantity of seed that he thought to be of a perennial variety. Because of exceptionally dry fall weather, the seed did not grow well and he became discouraged in trying to grow a new variety.

In the fall of 1931, Frank Kropf of Harrisburg planted a small field with perennial ryegrass seed imported by Jenks-White Seed Co. from New Zealand. This growing attempt also proved unsuccessful. The following year, 1932, Mr. Kropf imported 100 pounds of seed from New Zealand and planted ten acres of land. This planting stock came from a New Zealand field that had produced seed continually for more than 40 years.

The winter of 1932-33 was exceptionally cold with the temperature falling to nine degrees above zero in December, 19 above in January, and 11 degrees above in February. As a result of the severe cold weather, almost all of the ryegrass in the Willamette Valley froze out, but not the ten acres of New Zealand certified seed. This was the beginning of commercial perennial ryegrass seed production in the Willamette Valley. Later, Linn perennial ryegrass will be described and it will be seen why the above information is important.

From plantings in 1932 by Mr. Kropf, and the plantings of Smucker Bros., Paul Jensen, Wm. Windell, and Carl Keen, all of Harrisburg, and Levi Kropf and Wilbur Evans of Halsey the following year the acreage jumped to 4,798 acres by 1939.

As perennial ryegrass seed production began to expand in the early 1930s, the fluorescence test became available. This test has been used as a tool for making approximate determinations of purity in perennial ryegrass seeds in terms of mixture with common or annual ryegrass. This was necessary because both seeds are identical in appearance.

The Gentner Fluorescence Test to distinguish between *Lolium perenne* and *Lolium multiflorum* was introduced in 1929. However, it was discovered that some seeds of *L. perenne* also showed fluorescence. In the beginning this was thought to be caused by cross pollination. Mr. Henry Rampton and others proved that some *L. perenne* may show varying percentages of fluorescence.

One of the biggest criticisms of Oregon-grown perennial ryegrass in the 1930s was that it sometimes contained as much as 80-90% annual ryegrass. By using the Gentner Fluorescence test Mr. Rampton*, in 1933 for a Masters thesis, determined that the perennial ryegrass being grown in Oregon contained five percent natural fluorescence. As a result, seed laboratories around the United States, including the Federal Laboratory, wrote into their procedures a 5% tolerance for all perennial ryegrass varieties. The formula was 1.05% X total ryegrass in purity X % non-fluorescence divided by germination %.

Although perennial ryegrass may have been certified as early as 1937, Oregon adopted a set of standards for certifying perennial ryegrass around 1941.

The fluorescence level determined the color of the certification tag.

Tag Color	Max. Fl.%
Blue tag	.5
Red tag	15
Yellow tag	.30

During the 1950s very little field inspection of perennial ryegrass was undertaken. The fields were planted with the very lowest fluorescence seed stock obtainable and harvested. The seed was cleaned and tested and if it met the certification standards the grower would send in his acreage fees and the blue tags were issued. By this time the different color of tags had been dropped. The certification of the variety was rather generic with the tags reading, "Oregon Perennial Ryegrass."

---

*Mr. Rampton was a USDA Seed Investigator stationed in Corvallis.

New varieties of perennial ryegrass were not very prevalent although a few did come into Oregon prior to 1960. These varieties were products of breeding programs from overseas. Several individuals from Oregon traveling in Europe in the 1950s were told that Oregon Perennial ryegrass was an "Ecotype" (a variety that has been formed by ecological conditions) and as such would not be eligible for sale into the European Common Market. There were several ecotype varieties of perennial ryegrass eligible for marketing throughout Europe. These varieties had been given names and were on a generation program. It was obvious that this was a necessary procedure if we intended to sell Oregon perennial overseas. Thus, in 1961 the variety Linn perennial ryegrass was accepted for Seed Certification.

LINN PERENNIAL RYEGRASS (A statement issued by OSC)

Linn perennial ryegrass was accepted for certification by the Oregon Seed Certification Board on February 6, 1961. Linn is an ecotype which has been grown by two elite seed producers continuously since 1931 in a well isolated environment. It was introduced into Oregon from New Zealand in 1928 and had been grown on one farm for forty years. Seed from this particular planting was used to establish the original planting on the two farms which have grown this particular strain since 1932. These growers have always grown certified seed, but it has only been recognized by the crop designation of "perennial ryegrass." This eco-type stock will henceforth be recognized as the variety Linn certified. Only registered seed can be used for producing certified seed. There will be the appropriate field inspection before it can qualify for certification. These field inspection requirements are listed as a part of the Oregon Certification Standards which are published each year.

"In the early thirties, when ryegrass seed was first produced for domestic and international commercial marketing by Oregon, the chief concern was to differentiate between common or Italian (*Lolium multiflorum*) and perennial or English (*Lolium perenne*). The fluorescence test was then made available so that quite accurate identification could be made between the two species by this seed testing laboratory technique. For the past 20 years, the field inspection of perennial ryegrass for certification has not been used. It has been quite evident to research workers and certification specialists that the field inspection phase of certification of perennial ryegrass should be renewed. Therefore, for some 4 or 5 years considerable attention has been given to ascertaining some of the better stocks being grown for seed production in Oregon. The result of this study and research on relative seed producing capabilities was the acceptance of the Linn ecotype stock for certification. For those who might read this statement in Western Europe or Great Britain, this stock is now being tested at several locations under the identification of Linn 1 and Linn 2. The numbers refer to two experimental stock lots which have been bulked for purposes of seed production since no significant difference was shown between their seed producing potential.

"There will be a gradual transition from the marketing of certified Oregon perennial ryegrass as "perennial ryegrass" to the marketing on a variety basis solely. A program will be put into effect immediately with reference to initiating the certification of the variety Linn and discontinuing of certification of Oregon "perennial ryegrass". Both will be certified simultaneously through 1966.

"The following steps will be taken to effect this transition:

Perennial ryegrass

1. Oregon 'perennial ryegrass' cannot be planted for certification as 'perennial ryegrass' after 1961.

2. The 1966 crop of Oregon 'perennial ryegrass:' will be the last crop to be certified as 'perennial ryegrass.'

Linn

1. The 1961 crop grown by William Glaser and the Wheeler Bros. will be declared as registered class of seed provided it meets the necessary field inspection requirements for certification in this class.

2. Any plantings made in 1961 which trace to the established source (Wheeler Bros. and William Glaser) could be eligible for certification as Linn perennial ryegrass.

3. After 1961 only registered seed of the Linn variety of perennial ryegrass may be used to establish fields to produce certified seed."

This statement was prepared by Dr. J. Ritchie Cowan, Head of the Farm Crops Department at Oregon State College and Chairman of the Oregon Seed Certification Board.

In a newspaper article of June 3, 1971, Dr. Orvid Lee, U.S.D.A., was mentioned for his developmental work on the use of charcoal for banding and other sophisticated measure for establishing clean perennial grass seed crops. Prior to this technique all seed growers could do was plant perennial ryegrass seed fields with low fluorescence seed stock and harvest the first year about a 50-50 crop. Half being annual ryegrass and half perennial. By the second and third years the field, through the use of appropriate herbicides, would be cleaned sufficiently to meet the Seed Certification requirements. With the new methods promoted by Dr. Lee many perennial ryegrass fields were able to meet the certified requirements the first year. This was not true 100% of the time as the success was based upon the care and timeliness of the chemical application. However, there were many success stories.

Charcoal band planting

With the advent of new varieties in the 1960s, the fluorescence issue once again raised its ugly head. The European varieties carried no fluorescence value as the Europeans placed very little faith in this test. As new varieties came to the Seed Certification Board* for approval into the Oregon program the question as to how to handle them arose.

*Advisory Board established by the Dean of AGR OSU.

Basically, most of the new varieties were handled and listed in the Certification Handbook similarly to Linn perennial ryegrass. They were given the 5% tolerance and in some cases the plant breeder thought that his variety might even have as much as 3% more beyond the 5%. If the variety had no fluorescence to begin with, some varieties were being certified with as much as 8% annual or hybrid ryegrass in the sack of seed. During the early 1970s several varieties of fine leafed perennial ryegrass were introduced to the public. These became very popular turf-type perennials. Because these varieties were dark green, narrow-leafed and shorter growing any annual ryegrass in the lawn or sod stood out as a sore thumb. The sod growers were the first to complain and start litigation.

The Manhattan perennial ryegrass breeder and the Growers Association asked for the 3% tolerance instead of the 8%. The matter came to the attention of the administrator of the Federal Seed Act, Clyde Edwards. Basically, Mr. Edwards changed his definition of "off-type." The Federal Seed Act defines off-type as a plant or seed which deviates in one or more characteristics from that which has been described by the plant breeder as being usual for the strain or variety. Mr. Edwards took the position that off-type should be defined as any seed or plant not a part of the variety in that it deviates in one or more characteristics from the variety as described and may include: a seed or plant resulting from cross pollination by another kind or variety; a seed or plant resulting from uncontrolled self pollination during the production of hybrid seed; or segregates from any of the above.

Immediately, two camps were formed. One side, mainly from the Seed Trade, wanted the status quo kept. They could see that if stricter rules were put in place the volume of seed meeting the blue tag seed certification requirements might be drastically reduced. On the other hand, those who were truly trying to meet a quality conscious market knew they had a vast market and wanted to capitalize on the quality issue.

At this juncture Oregon State University organized the Ryegrass Symposium. It was apparent that few people knew very much about all of the ramifications of the fluorescence test and how it affected them personally, as growers, sellers or plant breeders. OSU decided it was time for a Ryegrass Symposium to be held April 24 and 25, 1979.

Over 300 people registered. On the program were such people as Mr. Arne Wold, 1st Vice President of the International Seed Testing Association, Mr. Henry Rampton explaining his research that allowed a 5% tolerance, Mr. Bob Bronson, a ryegrass grower explaining his cultural practices, a ryegrass breeder, Mr. Richard Hurley, Mr. Jost Barenbrug, a European seedsman and of course, Mr. Clyde Edwards, administrator of the Federal Seed Act. Mr. Edward's surprise announcement that hence-forth the 5% tolerance would be done away with, except for Linn perennial ryegrass, stunned the audience. Mr. Edwards based his edict on the definition of what constituted a variety and thus in the eyes of the Federal Seed Act any seed lot with more than 5% fluorescence had to be labelled as a mixture. Immediately the American Seed Trade Association branded the whole affair a crisis. OSU was accused of being in cahoots with Mr. Edwards and the fight was on. The American Seed Trade Association began to make moves to have Clyde Edwards fired. During the A.S.T.A. annual meeting held in Washington, D.C. a contingent met with the Secretary of Agriculture to no avail. Even one prominent seedsman in Oregon wrote the Secretary of Agriculture with a copy to President Nixon.

Because the 1979 Handbook had already gone to press, it was too late to alter the fluorescence allowances listed for the different varieties of perennial and annual ryegrasses. As a result, the administrator of the Federal Seed Act wrote to the Chairman of the Certification Board to say that our certification standards were out of compliance.

Eventually the office of the Federal Seed Act administrator issued some written rules listing certain varieties and how they were to be tested. Mr. Edwards did take early retirement. And, even though the stringent standards were put in place the market bounced back very quickly. The seed growers rose to the task and qualified their seed by doing more spot spraying. The seed industry overall benefitted by what took place.

The final resolution the industry lived with for over 10 years was: All varieties of perennial ryegrass were assumed to be non-fluorescing and annuals were 100% fluorescing. There were some known exceptions, i.e.: Linn perennial ryegrass, Magnolia, and Pennant. Plant breeders believing their varieties could not meet this requirement were to submit test data to the Grass Variety Review Board of the Association of Official Seed Certifying Agencies requesting a variance.

GULF ANNUAL RYRGRASS

Gulf Annual Ryegrass was first mentioned publicly in 1959. It was tested for seed yield and maturity date by Hank Rampton. During the 1959 Seed Growers League this variety was brought to the growers attention. By 1961 there were 508 acres certified. The common complaint from growers and dealers was that they could not find land that would meet the requirement of, "five years out of any ryegrass." The question was asked if a field was planted with Gulf annual ryegrass for five years could it be planted on the sixth year with Registered seed and would it be eligible for certification thereafter? The Seed Certification office discussed this and agreed that if Registered seed was planted for the previous five years, then the field history requirement would be satisfied. To a lesser extent this same procedure was used to establish history for Florida Rust Resistant annual ryegrass fields.

Gulf annual ryegrass was bred and developed in Texas. Since the variety yielded as well as Oregon annual and there was a price advantage, many growers converted their fields to Gulf. Very little acreage was certified. Most Gulf was planted on land that had been growing Oregon annual for years and years and as such produced a mixture of annuals. Complaints were heard from Texas, "Most of the Oregon grown Gulf annual ryegrass was losing its Rust Resistance." When questioned, Texas growers admitted planting uncertified seed. The matter was dropped.

Curiously the Foundation Seed stock arriving from Texas carried a percentage or two of non-fluorescence. Texas Foundation Seed officials did not use the fluorescence test and were at a loss to explain the non-fluorescence. If the fluorescence gene is dominate, this phenomenon should not have happened.

TALL FESCUE

In his report to the 1969 Seed Growers League, Henry Rampton mentions that: "In the agriculture report for 1905, the value of tall fescue as a pasture plant was recognized. I wonder why it was forgotten during the next thirty years. Perhaps an authentic and adequate seed supply and enthusiastic promoters were lacking."

During 1918 Harry Schoth of the USDA Agriculture Research Service, obtained several different lines of tall fescue from the station at Pullman, Washington. The variety "Alta" evolved as an ecotype selection from these earlier plantings. In the winter of 1922-23 the plantings suffered severe winter killing. The surviving plants were put together and became the source of "Alta."

In 1931 the Kentucky Agricultural Experiment Station had selected an ecotype of tall fescue from a mountain farm in Menifel County, Kentucky. This variety was released by the Kentucky station in 1942 as Kentucky 31. For years the administrators of the Federal Seed Act stated there was no difference between Alta and Kentucky 31 and regarded them as the same variety. On September 1, 1974, USDA Seed Regulatory announced that the two varieties would be recognized as separate varieties. There was no explanation given for their sudden about face.

Alta Tall Fescue

Floyd Mullen in his book, "The Land of Linn" states: "Alta fescue, a forage grass and seed crop, was started in 1935 when Frank Rohwein of the Jordan community harvested seed from a one acre planting. The next year Wilbur Evans, Halsey, planted a small acreage for seed purposes."*

*I believe that early planting of Wilbur Evans was just across the Peoria Road from the present Mennonite grade school.

The Extension Circular No. 334, "Production and Income Statistics for Specialist Farm Products" reports 70 acres of tall fescue in 1938. These were grown in Clackamas, Linn, Marion, Washington, and Union counties. By 1939 this acreage had grown to 150 acres yielding 23,500 pounds of seed which was sold by the growers for $.34 per pound.

So far as we in the West were concerned there were no new varieties of Tall fescue until "Fawn" tall fescue was released in 1964 by Oregon State University.

ORCHARDGRASS

In 1935 during a seed production class G.R. Hyslop told the students that "orchardgrass" is probably the most commonly used perennial grass in western Oregon. He further stated that mixtures sown on burned-over land usually contain from 12 to 25% orchardgrass. Several railroad car loads were being sold in the Northwest each year and very little seed is produced in the Northwest.

The first acreage certified was in 1952. It was an eight acre field of S-143 orchardgrass. The variety Akarora was soon added. Several years later the variety Potomac was released and rapidly became the favorite variety to grow.

The two early varieties, Akarora and S-143, were foreign varieties that had been grown and further selections made on them by the Soil Conservation Service at Mt. Pleasanton, California. Instead of renaming their new selections they simply released them with the old names. The Europeans took exception to this and asked us to stop certifying these two varieties by these names. This was an unfortunate set of circumstances as the varieties were well suited to the western states and were grown uncertified for years.

BENTGRASS SEED PRODUCTION

The November 1930 issue of "The Bulletin" of the U.S. Golf Association Green Section carried two articles on bentgrass seed production in the Northwest. In one article, H.A. Schoth, U.S.D.A. stationed in Corvallis, outlines the history of the bentgrass production in Oregon and Washington in the early 1920s.

"No literature on bentgrass seed production in the Pacific Northwest is complete without reference being made to the late Dr. C.V. Piper. Dr. Piper, for many years intimately associated with plant life in the Pacific Northwest, identified many of the bentgrasses growing in this territory and on numerous occasions expressed the belief that some day it would become a large seed producing section. He used his influence whenever possible to create interest in bent seed producing possibilities.

His predictions have been fully substantiated, and already more bentgrass seed is being produced in this territory than in all the remainder of the United States.

The history of bentgrass seed production in the Pacific Northwest began in 1923 with the determination by Roland McKee, of the Bureau of Plant Industry, United States Department of Agriculture, and the writer, that the dense, low-growing, sod-forming grass growing in the Coos Bay, Oregon, region, and known locally as Bermuda grass, was in reality creeping bent. As a result of this determination the foundation for the present Pacific Coast bentgrass seed industry was laid.

Upon Mr. McKee's return to Washington, D.C., this information was transmitted to Lyman Carrier, at that time in the office of forage crops, Bureau of Plant Industry, United States Department of Agriculture. Mr. Carrier was very much interested, and during the early spring of 1924 made a trip in the company with the writer, Harry Schoth, to the Coos Bay region for the purpose of investigation and ascertaining the possibilities of future bentgrass seed production. He was very much surprised to find such large areas of practically pure stands of bentgrass. A few months later, during that same year, Mr. Carrier became actively interested in harvesting and preparing for market the first bentgrass grown on the Pacific coast. The harvesting of bentgrass was begun two years later in the lower Columbia River section, and new areas are developing with each succeeding year.

From that comparatively small beginning the industry has developed very rapidly and the territory now producing bentgrass seed extends from southwestern Oregon to the Canadian line, mostly west of the Cascade Mountains. The largest bentgrass seed producing areas are the Coquille Valley, the lower Columbia River districts in Oregon, and the Puget Sound district in Washington. Smaller producing areas are at Reedsport and Gardiner near the mouth of the Umpqua River, at Cushman near the mouth of the Siuslaw River, in the vicinity of Yoncalla, in the Willamette Valley, near Klamath Falls in Oregon, and in various sections of western Washington chiefly in the vicinity of Chehalis and Raymond.

Practically all of the stands now being harvested for seed are natural and many of these have been in existence for many years. The only entirely artificially seeded fields of bentgrass now being harvested for seed are in Klamath County, Oregon. These plantings consist of approximately 70 acres and produced their first seed crops in 1930.

For the most part the land on which these bentgrasses are growing is moist, of very low elevation and quite often subject to overflow, being in many cases under almost tidewater conditions. The regions of highest elevation now producing bentgrass seed commercially are in the vicinity of Yoncalla and Klamath Falls, Oregon. The elevation at Yoncalla is approximately 356 feet and that at Klamath Falls approximately 4,105 feet."

Dr. D.D. Hill, a former OSU Crop Science Department Chairman, was able to shed a little light on the varieties Seaside and Astoria bentgrasses:

"In 1920, the only bentgrass seed available in the United States was an imported German mixed bent which sold for the then astronomical price of five dollars per pound. An agrostologist of the U.S. Forage Office by the name of Lyman Carrier, made periodic trips to the west coast. During one of these trips he discovered a bentgrass being grown for hay by the dairymen of Coos County. This strain was remarkably pure, but no one had ever thought of saving the seed, as it volunteered spontaneously in the wet meadows of Coos County.

Mr. Carrier resigned his position with the Department of Agriculture, moved to Oregon, and formed a company known as Cocoos. He would go to a farm with threshing equipment, pay the farmer fifteen dollars a ton for his hay, thresh it, and then give the straw back to the farmers. This was a good deal for the farmer, but when Professor Hyslop became aware of this practice, he began to find out how much seed was being produced and how much was being paid for it. His immediate conclusion was that the local growers were getting the short end of the stick. He convinced a number of growers to let him certify the seed to be Seaside Bentgrass. Actually, there was no legal authority for him to do this. Certification was done by the Oregon Agricultural College. This proved to be so rewarding to the growers that soon most of them were applying for certification instead of selling their hay to Mr. Carrier. Mr. Carrier refused to handle seed of the certified seed growers, who were then forced to find other market outlets. I don't know what these outlets were, but apparently they were adequate as there appeared to be no problem of disposing of the seed. Mr. Carrier wrote a very strong letter to President (OSU) Kerr, asking that Prof. Hyslop be fired. Dr. Kerr simply sent the letter over to Prof. Hyslop. I found this letter in Professor's desk after his death in 1943. Cocoos bent eventually disappeared from the market."

ASTORIA BENTGRASS

When Albert Engbretson was Superintendent of the Astoria branch OSC Experiment Station, he discovered a bent growing in the wet meadows around Astoria. With the help of Prof. Hyslop and Miss Grace Cole, U.S. seed analyst at Corvallis, it was determined that the bent from Astoria was distinctly different from that grown in Coos County. Shortly thereafter, Mr. Engbretson resigned his position with the experiment station and formed the Engbretson Seed Company. I am not sure of the date, but I think it was around 1925 or 1926. In contrast to Mr. Carrier, he used the certification service from the beginning. Later on, however, when he felt that his company was well enough established to sell seed on the company reputation, he sent a carload of seed east without certification, only to find that he was unable to sell it. It was necessary to ship the carload of seed back to Oregon to complete the certification process.

As mentioned, the variety Seaside was originally discovered near Coos Bay in 1924, and as George Hyslop explained, was first certified along the coast. Harold Finnell told of accompanying Mr. Hyslop on weekends.

The two men would leave Corvallis for "Cushman" by train after the last class on Friday afternoons. Cushman being located near the mouth of the Siuslaw River. There they would be met by bentgrass growers and taken by boat to the farms. Mr. Hyslop explains how bentgrass was examined and certified during the weekend. As soon as the seed sacks were tagged and sealed Sunday afternoon the two men would catch the train and be ready for class Monday morning.

From the early beginnings the three different varieties and/or species soon found areas of the state where they produced the most and cleanest seed. For example, Seaside creeping bentgrass was grown along coastal streams which overflow. The water may stand throughout the rainy season. This treatment kills out most weeds and other grasses leaving remarkably pure stands of bentgrass. The grass spreads quickly by runners and so fills up vacant spaces. Any weeds which come in are killed by the next overflow period.

In Klamath County the Geary Brothers and George Stevenson tried producing Seaside bentgrass on reclaimed marsh land. About 5000 pounds were produced in 1930. The Geary Brothers found that they could duplicate the coastal conditions by using the reclaimed marsh land and flood their fields in a similar manner as the coastal areas. For nearly forty years there was 1,000 acres of Seaside bentgrass grown on the Geary Ranch.

Mr. Shirl Jackson in southern Benton County and northern Lane County would produce Seaside in a similar fashion always flooding the fields. Shirl originally obtained his seed from the Cushman area and always reseeded fields with his own seed.

Commercial seed production of Astoria Bentgrass was primarily centered in the Lower Columbia River Valley. Developed primarily through the efforts of Mr. A.E. Engbretson, the first seed was harvested in 1929. The main areas where Astoria was grown were in and around Clatskanie and Astoria, most generally on marshy, boggy ground or peat soil that could not be burned after harvest. As a result, the Astoria bentgrass production was driven out of these areas and into the Willamette Valley because the ergot and nematodes could not be controlled by burning. Astoria never did achieve the prominence that Seaside and Highland bentgrasses were able to achieve. More on this subject later.

HIGHLAND BENTGRASS

Highland bentgrass was first recognized as a distinct type in the Umpqua Valley near Yoncalla, Oregon in 1926 by George Hyslop and Harry A. Schoth.

The prevailing feelings are that this bentgrass probably came over to the United States from Germany or Northern Europe as packing material in crates. The story is told of large crates being opened in Yoncalla, Oregon. The straw inside the crates used for cushioning the equipment was kicked out the back door of the store, and eventually Highland bentgrass was identified.

The first seed was harvested near Yoncalla in 1927 from a natural stand of some seven to eight acres. It is reported that this first production was under the supervision of the Jenks-White Seed Company, Salem, Oregon. Some four years later, 1930, this type of bentgrass was found on a small area in the Willamette Valley near Hubbard, Oregon. In 1934 it was given seed certification status and the first certified seed was harvested that year from the farms of Mr. A.R. Colman, Hubbard, Oregon (44 acres), and acreage at Halsey, Oregon, by LeRoy Nicewood. Farmers did not plant Highland bentgrass to establish it for seed production. In fact in the hill soil section of Marion County (Silverton to Stayton) farmers fought it as a weed for a few years before they started cultivating it. The heaviest establishment was on some of the more run-down soils. It took a few years to find out how to handle the grass most

George Dewey in Highland Bentgrass, mid 1950s

advantageously. But once it was worked out by the farmers the acreage increased by leaps and bounds. By 1937, Savage, Brownell, and Darby were growing, harvesting, and threshing Highland near the Victor Point School. Savage had a cleaning plant for putting the seed in condition for sale and may have been the first commercial seed cleaner in the area.

By 1954 the acreage had increased to 6,182 and doubled by 1957 to 12,641. Much of this seed was being sold overseas.

During the mid-'60s those who traveled overseas were told that Highland bentgrass was going to lose its status if it was not put on a generation system and entered in the turf trials for comparison. Thus, Dr. Rodney Frakes, 1960-1989, OSU Grass Breeder, went to a number of certified seed fields in the Willamette Valley and selected several plants that formed the Foundation plantings. The seed that was harvested from there was planted to produce all the Registered seed needed. By 1972 all fields planted had to be with Registered seed. As far as certified seed was concerned, the 1978 crop year was the final year for those fields that had not been planted with seed stock.

HIGHLAND BENTGRASS CHARACTERISTICS

Highland bent can be credited as being a robust grass, but not in the sense of having large massive stems, leaves, crowns or roots and unusual stamina, but in the sense of its ability to make good growth under a wide range of climatic and soil conditions.

Highland bent has a very extensive relatively fine root system with fairly deep penetration. It is a good sod former and spreads quite rapidly by rhizomes and stolons. Both rhizomes and stolons root at the nodes and consequently with the two methods of vegetative plant reproduction any vacant spots, regardless of why, are soon covered or replaced by vegetation. It is this situation that on occasion adds to its durability and persistence.

Highland bent under most favorable growing conditions will reach a height of from 14 to 16 inches. The stems are fine and have a tendency to lodge. The heads are of the open panicle type and quite large. The seeds are small with more than two million per pound.

Highland bentgrass, as all of the bents do, matures one month later than the other cool season grasses grown for seed in the Willamette

Valley. Because of this, Highland suffered very little from competition from other grass species. The velvetgrasses were a problem until a complete herbicide program was worked out. Additionally, "Red Top" or *Agrostis alba* continued to show up in the seed fields and the commercial seed. Gradually these two plants cross pollinated and the hybrids began to appear in these fields. Eventually the red top disappeared completely.

All through the history of the bentgrasses there has been confusion over the exact species. There seems to be ample evidence to show that Highland should never have been called a Colonial bentgrass, but this mistake has never been corrected. This was due mainly to the fact that Highland was known throughout the world as a colonial bentgrass and no one wanted to tamper with this designation.

As the bentgrasses in Oregon evolved, the Seaside Creeping bentgrass was a vigorous creeper, spreading only by stolons. Astoria colonial bentgrass was a much less vigorous creeper and only crept with weak stolons. Highland colonial bentgrass, on the other hand, spread by vigorous rhizomes and stolons.

OTHER BENTGRASS TYPES AND VARIETIES

Over the years there have been numerous attempts to grow other varieties and species of bentgrasses. Kingstown velvet bentgrass was tried but never caught on. Some of the varieties used for golf courses, on the other hand, have been very popular, i.e., Penncross, Penneagle, and Pennlinks. These varieties were developed at Penn State and reproduced at the seed stock level vegetatively and are very popular for golf greens. Several varieties were sent from overseas to be planted and reproduced with little success. Some of the seed stock arrived mixed while others were planted on ground that contained Highland bentgrass and were soon contaminated.

Penncross creeping bentgrass was released from Penn State in the mid to late '50s. It was a product of Professor H.B. Musser's breeding program. It was a unique variety in that there were three lines or clones, which were propagated vegetatively. The seed growers in Oregon would receive stolons of the three clones in sacks and plant these in three distinct beds. As the beds grew and matured portions would be dug and shredded and the stolons would be planted by using a strawberry planter all in separate rows. A seed field would consist of rows of the three. The three clones were allowed to freely interpollinate and the seed was taken from the field.

During the mid '60s on a very hush, hush note we were told that some of the older beds had plants in them with stripe smut. Had this disease been allowed to continue it would have been the end of Penncross production. Since it was very devastating to lawns on the East Coast it might have marked the end of all lawn grass seed production in the west. Fortunately, cultural and chemical practices were developed that corrected the problem.

In addition, around this same time rumors started circulating that Oregon was certifying Penncross seed that came from seed fields established not with clones but with seed. Several golf course magazines carried articles complaining about the quality of Penncross seed grown in Oregon. At this juncture a few seed growers (Bill Rose, Lester Estergard, Don Hector, and seedsman Dick Bailey to mention a few) formed the Penncross Bentgrass Growers Association. The association has become very successful over the years. It promoted the use of Penncross for golf greens throughout the world. The association obtained exclusive rights to the clones for its members and fought the battle of name retention only when the three clones were planted in the approved manner. Thus, Certification became very important to the growers.

At times the price a grower received for his seed was $3.00 per pound for certified seed versus $0.80 for uncertified. No other grass species has had as much riding on the field inspection outcome.

RED FESCUES (FINE FESCUE) CHEWINGS AND CREEPING

The first commercial production of chewings fescue in Oregon was in 1933. Prior to this, considerable seed was being imported from New Zealand. Chewings was named after Sr. Thomas Chewings, a New Zealander who developed the strain. Usually the imported seed arrived low in germination or soon lost germination.

Extension Circular #318, September, 1938, states that there were 50 acres of Chewings fescue harvested in 1936. By 1939 that had grown to 925 acres and to 7,260 acres by 1947.

Chewings fescue and creeping red fescue both belong to the same species (*Festuca rubra*). Chewings as a sub-species is a tufted plant with no creeping habit. Creeping red fescue does have strong rhizomes and thus creeping habit. Both have fine blades and make a desirable lawn with shade tolerance.

Creeping red fescue farm, 1939

Research done at Rutgers seemed to prove that the two types had different chromosome numbers. However, the world collection of Festuca rubra varieties indicated considerable overlap of chromosome numbers within the types. Thus, both of the fine fescue types had to be grown isolated from each other.

A report by the O.S.C. News Service dated October, 1940, stated, "Three men in western Oregon harvested enough seed this year of creeping red fescue to provide a combined supply of about 8,000 pounds according to Harry A. Schoth federal agronomist at Oregon State College, who first introduced the grass on the station test grounds in 1929." The three growers were; Gilmore Hector, Albany, Oscar Loe, Silverton, and Harry Riches, Salem, a combined acreage of 20 acres. By 1946, there were 2,275 acres harvested.

The first varieties were Oregon Creeping Red Fescue, Rainier Red Fescue, and Illahee Red Fescue. Pennlawn Creeping Red Fescue, a release from Penn State, soon took over as the primary variety.

The Extension Certification annual reports show that 52 acres of Illahee were certified in 1945 and 15 acres of Rainier in 1946. Although Chewings fescue had been grown since 1933, it had not been certified. In the 1960s, Oregon State University released a variety called "Cascade" Chewings fescue.

HARD FESCUE, IDAHO FESCUE

Idaho fescue was being promoted for seed production in the 1940s.

The Soil Conservation Service in Pullman, Washington released a variety of hard fescue for dryland pastures and soil protection called Durar (P2517). This variety had been collected from an old planting on the Eastern Oregon Branch Experiment Station, Union, Oregon, in 1934. It was ultimately released by the S.C.S. in 1949.

A hard fescue variety, first certified in 1974 as C-26, was a foreign variety originally released as Biljart hard fescue. The administrator of the Federal Seed Act ruled that the name could not be changed when brought to the U.S. Thus, in 1975 the company had to relabel and certify as Biljart.

KENTUCKY BLUEGRASS, ROUGH BLUEGRASS, AND BIG BLUEGRASS

Hank Rampton in his 1969 speech to the Oregon Seed League reported "that the first known harvest of Oregon grown Kentucky bluegrass seed occurred in Klamath County in about 1926. He states that he saw some of Ed Geary's seed in 1931 and was amazed at its large size and plumpness." Extension Circular #318, September, 1938 shows that there were about 700 acres of Kentucky bluegrass grown in 1936 -- all in Klamath County. The expectation was that the yield would be 1,000 pounds per acre.

It is my understanding that Ed Geary used considerable bluegrass in his pasture land. He harvested seed fields and always cleaned his own seed. The seed that he kept for his own use was always the biggest and plumpest seed of the highest bushel weight, thus, developing a variety of Geary Kentucky Bluegrass. This variety was traded by the seed industry as late as the 1960s.

MERION KENTUCKY BLUEGRASS

The first new improved dwarf Kentucky bluegrass was Merion. The first plantings in Oregon were by rhizomes in Klamath County about 1940. Merion when first released was labeled "B27". The certification records show 32 acres certified in 1948 and 41 acres in 1949. In 1952, the acreage jumped to 379 acres. This was due to the efforts of Otto Bohnert in Central Point, Oregon. Otto had been given a few grams of Breeder seed. By growing this seed in flats in his greenhouse he was able to obtain enough material to plant a small Breeders block. He soon had sufficient Foundation seed for himself and the Central Oregon growers.

Merion Kentucky bluegrass was released through Pennsylvania Agriculture Experiment Station who had the responsibility of maintaining the breeders seed. Subsequently, some research data from Dr. Joe Duich of Penn State showed Merion to be highly apomictic. Results of first and second generation progeny tests indicate the range of aberrancy averages from 2.8% to 4.2%. Aberrants strong enough to compete with the Merion strain occurred in a relatively low frequency. Approximately 19% of the total aberrants, or approximately 0.76% of the progeny were found to be plus or equal.

This is mentioned here because there were many interpretations of this work. Seemingly everyone was willing to tolerate a high degree of contamination in the variety Merion, and thus set the stage for allowing larger quantities of contamination in certified seed of bluegrasses than would have normally been allowed.

As long as Otto Bohnert grew and produced the foundation plantings of Merion he would scour the fields every year for these so-called off-types. The Seed Certification standards allowed up to 3% off-type Kentucky bluegrass in the certified class of seed. As the acreage shifted from Central Point to Madras, the amount of contamination increased. At one point in the late 1950s and early 1960s very little Merion was able to meet the 3% off-type level and very little received the blue certification label, because of a high percentage of common Kentucky bluegrass.

At this stage of production, the seed laboratories had no trouble distinguishing Merion from Common Kentucky bluegrass. What was perceived to be a small amount of contaminant in the variety was really Common Kentucky bluegrass. The Bureau of Reclamation had seeded Common Kentucky along the main canal when they completed the new irrigation district in Madras. This common seed crept into the seed fields. Looking back now, this contamination should not have been tolerated in certification field inspections, as allowing this set the stage for the next round of bluegrass growing in Madras.

NEWPORT KENTUCKY BLUEGRASS

The Ferry-Morse Seed Co. through Buchanen-Cellers of McMinnville asked that C-1 Kentucky Bluegrass be certified in Oregon on a hush-hush basis. This occurred in the late 1950s. Next year, Harold Finnell* was asked to certify the C-1 variety as C-1 Newport Kentucky Bluegrass which was done. Also, during this year we were asked to look at a field of Newport Kentucky Bluegrass that was to produce foundation seed. The original Breeders seed was released by the Plant Materials Center, S.C.S., Pullman, Washington. A problem was developing. Ferry Morse admitted that their C-1 or Altra C-1 had come from the same source as Newport, but their plant breeders had made further selections on the variety and thus could call their variety by a different name.

An urgent and very tense Seed Certification Board meeting was held. Dean Earl Price, O.S.C. Agriculture Dean chaired the meeting. There were several lawyers present and many on-lookers. Newport originated as a single apomictic plant selected from a maritime race on the coastal bluffs at Newport, Oregon, by M.E. Lawrence. The plant was further selected by the Plant Material Center staff, S.C.S., Pullman, Washington, Jens Clausen, Carnegie Institution of Washington, and Stanford University, Stanford, California. The strain accession C1W 4466-1 and P-13821 was used in Carnegie Institution of Washington hybrid bluegrass studies.

If minutes of this board meeting could be found, perhaps Ferry-Morse explained how and where they obtained seed and what further selective work was done. The on-lookers were asked to leave. The upshot of the meeting was that Dr. Rodney Frakes, O.S.U. plant breeder, was to conduct research and report his findings. This he did in 1961-62.

---

*Harold Finnell was the Seed Certification Project Leader, OSC, 1935-1963.

His conclusion was that there was a significant difference in the two seed sources. Nonetheless, the whole world was not convinced and the C-1 variety disappeared very quickly.

Newport bluegrass was not dwarf or low growing. It was a tall, very uniform distinct variety. thousands of acres were grown mostly in the Madras area. Much of the Newport was grown on Merion bluegrass ground. It turns out that the soils in Madras irrigation districts were ideal for storing viable seed. It was discovered later that bluegrass seed could live in the soil for at least seven years and retain a high degree of germination. The older Newport fields started exhibiting contamination. Several fields producing foundation and registered seed were never rogued for off-type. It was felt for years that the seed analysis would identify contaminants (off-types). This was true if the contaminate was Merion, but if it was common Kentucky bluegrass it would not be distinguished. Unfortunately, all of these negatives did not come out for several years. By this time there were at least 50 new bluegrass varieties and the Madras district had established itself as being able to get a seed crop the first year. Thus, every new variety was planted in August or September for the production of seed stock the following year. This was on ground that had produced bluegrass at some time or another previously.

The growing of certified Kentucky bluegrass turned out to be most frustrating for everyone.

OTHER BLUEGRASS TYPES

The Pacific Slope Dairy Farmer magazine in 1929 printed comments about Hanley-Hoover's Winter Bluegrass. C.C. Hoover (Charley) of Medford, Oregon was promoting *Poa bulbosa* as a Winter bluegrass for pastures in Southern Oregon and Northern California. Charlie Hoover was a very good supporter of Oregon State College and a personal friend of Dr. D.D. Hill.

Before World War II, Oregon grew 300 acres of Bulbosa for seed.

A couple of varieties of rough stalk bluegrasses were grown in Oregon for certification. One variety came from overseas, the other was released by a private company who deliberately mixed it with Kentucky bluegrass when it was sold as a brand. So, they planted their variety on land in Madras that had just grown Kentucky varieties.

BIG BLUEGRASS

This species was grown for seed in Oregon during the 1940s as recommended by E.R. Jackman. Big bluegrass is seeded with crested wheatgrass in areas where rainfall is 18 inches or more. The variety Sherman Big Bluegrass was selected and released in 1945 by the S.C.S. Plant Materials Center, Pullman, Washington. The name Sherman was derived because it was collected from native vegetation near Moro, Sherman County, Oregon.

MISCELLANEOUS GRASSES:

TIMOTHY

Timothy is indigenous to Europe except for Turkey. It is also native of Northern Africa and a large portion of western Asia and Siberia. Timothy was first grown in Oregon for certification in 1950. Only fair seed yields could be obtained in the LaGrande area. Growing Timothy for seed was tried in the Willamette Valley but only small amounts of seed per acre were harvested. Canada and northern Minnesota seem to be able to produce good seed yields. The assumption was that the northern latitudes with cold winters are needed to produce quantities of seed.

BROMEGRASS (*B. inermis*)

Perhaps the same story is true for bromegrass. A field of Baylor bromegrass was established east of Harrisburg in 1967. The field was harvested during some dry weather around the fourth of July. The comment was "that it was just like harvesting feathers." The seed was light and shattered easily.

Manchar Smooth brome was tried on four acres as early as 1948.

Both Timothy and non-native Bromegrass grow well in pastures; neither do well in Oregon for seed production.

TALL OATGRASS

The only variety of Tall Oatgrass certified was called Tualatin. This variety was first selected by H.A. Schoth in 1930. It was later released cooperatively by the Oregon Agriculture Experiment Station and A.R.S. The seed when cleaned shed the outer chaff easily, thus, the seed was sold as groated. The germination would fall fairly quickly as a result of the groating. The dust generated during harvesting and cleaning was so disagreeable that production of tall oatgrass did not stay around for very long. There was five acres grown in Clackamas County in 1936. By 1946 the acreage had grown to 225. By 1949 there were 254 acres certified.

REED CANARY AND HARDINGGRASS

These two grasses grow naturally on low, wet ground, along streams, ditches, marshes and sloughs. Seed production in Oregon has been on a limited basis mainly in the Willamette Valley. Isolation was always a problem. As a result, some production was taken to the hillsides. These grasses tolerated drought conditions very well. In later years there was very little demand for the seed. During 1936, Oregon growers grew 75 acres of Reed Canarygrass. Fifty acres of the 75 were grown in Coos County. In 1948, seven acres of Reed Canary and 100 acres of Harding grass were certified. Forty acres of the two crops were reported in 1950 and then very occasional acreage thereafter.

Mr. E. A. Jackman reported that Reed Canarygrass is one of the few cultivated grasses that is native to the area.

WHEATGRASSES

There was a big flurry of excitement in the mid '30s over Crested Wheatgrass. The variety was touted for the dryland areas of eastern Oregon. A number of demonstration areas were established in the desert near Prineville, Burns, etc. The Extension Service polled surrounding states for seed sources. Some responded from as far away as Brookings, South Dakota.

The records show that Oregon was growing and harvesting as much as 2,400 acres on the average between 1936 and 1944. However, the acreage fell to 500 in 1945 and 1946.

In 1948 a small acreage (15) was certified and this was expanded to 160 acres in 1949. In the '50s there were several kinds of wheatgrasses released by the Soil Conservation Service. Oregon grew small acreages of these but found that it was easy to flood the market as the demand came and went quickly.

Harvesting Crested Wheatgrass

SUDANGRASS

Sudangrass has never been a big seed crop in Oregon. Nonetheless, it has been grown for seed for decades. This species is planted in late spring in anticipation of some moisture falling during the summer. Quite often Sudangrass is planted to provide some early sheep pasture on the valley floor. If the pasture is not needed and the weather remains dry in September a seed crop is taken. However, there have been years when a demand out of California existed and growers planted with the full intention of harvesting seed. There were 720 acres grown for seed in 1936 and 3,200 acres in 1946.

Sudangrass looks very similar to Johnsongrass, a noxious perennial weed found in California. When Sudangrass is growing along the highways it has been known to fool the experts. It should also be mentioned that after a heavy frost the prussic acid build-up renders Sudangrass poisonous to livestock.

Additional Comments

Before leaving the grasses, a couple of items took place that need to be mentioned.

Organization for Economic Cooperation and Development–OECD.

This organization was an outgrowth of the European Marshall Plan started after the Second World War. The OECD deals with many items not the least of which is agriculture and seed specifically. It may be looked upon as an International Seed Certification Scheme. Each nation has one vote. The organization meets once a year and is dominated by the European nations.

The OECD Seed Certification has been a good program for the Oregon seed growers. We began this program locally in 1963-64 in Polk County. There were 453 acres of ryegrass harvested and tagged in 1964. From this one large field Oregon went on in later years to grow and ship many tons overseas under the auspices of this program.

Several ryegrass varieties coming into Oregon for further multiplication were tetraploids. This meant that instead of 14 sets of chromosomes, they had been doubled to 28. This was good in that tetraploids – either annuals or perennials – could grow next to diploids without fear of cross pollination. However, reports started filtering back from The Netherlands OECD grow-outs, that Oregon's production of tetraploids were contaminated with diploids. This was a serious accusation and if true and allowed to continue would have jeopardized our involvement with the OECD program. Fortunately, the Oregon Seed Laboratory started counting chromosomes and perfected a quick method of identifying diploids. Thus, we insisted that that test had to be run before OECD tags were applied.

The contamination did not seem to be deliberate but was probably the result of a combined number of different areas, where contamination had crept in. There was no question but what it was there, it was pointed out to me very vividly on a visit to The Netherlands.

Chapter IV

Legume Seed Production in Oregon

RED CLOVER

Red clover was one of the earliest leguminous crops introduced in Oregon, being grown as early as 1854. U.S.D.A. Circular No. 28 issued April 28, 1909 makes this observation: "When first brought into cultivation, the soils of the Willamette Valley were friable, quite easily tilled, and productive. For forty or fifty years the cereal crops were grown almost exclusively. As the soil became infested with weeds, summer-fallowing became a common practice. This system of tillage and the continuing growth of cereal crops year after year depleted the soil of much of its vegetable matter and rendered it heavier, more lifeless and more difficult to work. Farmers turned their attention to the growing of clover. Seed production in a commercial way did not begin until 1901 or 1902. The usual yield of red clover was from 4 to 6 bushels of seed per acre."

The total production for Oregon remained small until 1921 when over 1.5 million pounds were produced. Seed production fluctuated widely after 1929 and reached a record high of more than 4 million pounds in 1957. Western Oregon produced 3/4 of the crop in 1957. Washington county was the leading county producing 28% of Oregon's 4 million pounds.

Dr. D.D. Hill reported that for many years Oregon red clover seed enjoyed a good market position, largely because of its appearance. Oregon's favorable harvest conditions resulted in a seed that was bright, a seed that *looked* good. Unfortunately, the red clover, grown for many years in the mild winter temperature of western Oregon, lacked the winter hardiness required in the Corn Belt states, the principal market area. Some experiment station tests in the Midwest confirmed this in the late 1920s.

At about this same time the Tennessee experiment station developed one of the first truly new varieties of red clover. This strain was bred to resist the red clover disease, anthracnose. The variety was known as Tennessee Anthracnose Resistant, or TAR. Also, during the winter of 1929-30, Prof. Hyslop obtained a supply of seed that had been grown on the Hyslop family farm in Ohio for many years. This seed was known as Ohio Winter Hardy.

Supplies of both the Ohio seed and TAR were planted in western Oregon in 1930 under a hastily contrived certification program. The rules required that the seed be drilled so that checks could be made for volunteer clover plants. Field history required that no clover be grown on fields that had a history of recent red clover. Fields were inspected both during the seedling and the crop years.

Clover production in the Madras area

This was undoubtedly the first time the technique of a "seedling inspection" was employed. This was necessary because of the hard seed content of red clover. One of the reasons Malheur County never became a large red clover seed producer was because of the hard seed. Most of the red clover fields were planted back to alfalfa and the red clover would volunteer in these. In fact, a few acres of Foundation Vernal were rejected for too much volunteer red clover in Malheur County during the 1950s.

During the mid-1940s Kentucky released a new variety of red clover by the name of Kenland. It was decided by Kentucky to use the National USDA Foundation Seed Project to increase this variety. Oregon was contacted and asked if there was sufficient "red clover" free land in which to grow Foundation Kenland. The Madras irrigation project had just opened. Conditions turned out to be just perfect. Yields were outstanding. Central Oregon was suddenly on the seed production map.

To my knowledge we were never able to duplicate those yields again either at Madras or anywhere else in Oregon. The suspicion is that at first in Madras the land had only grown cereals, never red clover, so there were no viruses present.

Kenland when released by Kentucky was on a four-generation system. After several years of growing Kenland in Oregon and Washington with a registered class, it was discovered that the fourth generation was one too many as the genetics became altered.

Subsequently, recent Kentucky red clover releases have stipulated that seed cannot be taken the same summer that it was planted. In other words, the plants had to go through a winter before seed could be taken. Also, in an attempt to maintain genetic purity, seed fields were limited to two seed crops. Part of the reason for this was that during harvest the seed shattered and volunteered the following year resulting in multiple generations.

The average acreage of red clover grown for seed between 1936-39 was 20,400. This fell to 11,700 between 1940-44 and jumped to 20,600 acres in 1947. Much of the production has been uncertified especially acres in Washington and Yamhill county which did not qualify.

During the late '40s and the '50s a common practice in planting red clover was to hand-seed red clover with a horn (two- to three-foot metal cone-shaped device) in January or February onto established fields of winter grain. The freezing and thawing during the late winter early spring would break the hard seed coat and help establish the seed in the soil. Obviously, seedling inspections were difficult as there were no rows, but if red clover was present it was always up and well established ahead of the new seedlings.

Tetraploid varieties of red clover were introduced in Oregon from Sweden. Seed yields were low as the bees were hesitant to pollinate the big flowers. Bumble bees were the only bees large enough to accomplish this task.

ALSIKE CLOVER

Oregon has been an important producer of Alsike clover seed since 1930, when it ranked sixth in the nation. Oregon continued to expand production and for several years was the most important Alsike seed producing state. Oregon ranked first in 1956, second in 1957, and third in 1958. Most of the seed production was from east of the mountains. In 1939, Oregon produced seed from 20,000 acres, 7,400 of these were in Klamath county. Additionally, Klamath was the leading Alsike seed producing county in the nation in the 1954, producing close to a fifth of the nation's total.

Harvesting Alsike clover

None of the Alsike seed was certified presumably because there were no varieties. However, all field inspectors had to be trained to identify Alsike in Ladino and red clover seed fields. It was definitely a contaminant. Alsike seed contains hard seed that lives in the ground for years and years. A good example of this is growing white clover on land that had grown only ryegrass since having grown Alsike some twenty years in the past. Alsike showed up in the fields after 20 years.

LADINO CLOVER

Although records show that there were over 1,200 acres of Ladino grown for seed in 1936, it was not until 1949-52 that Ladino clover seed production really got underway. The original plantings were in Jackson and Josephine counties. However, with the new irrigation districts in Jefferson county, the production quickly shifted to the Madras area. In 1949 there were 10,000 acres of Ladino grown for seed and this doubled to 20,000 the following year. The seed certification office decided to place a person full time at Madras to look at all of the Ladino acreage in June, July, and August. Unfortunately, California discovered that they had some cheap land that would produce Ladino clover seed and the Oregon acreage dropped as quickly as it rose.

I'm not sure of the origin of the Ladino. At first it was grown without a generation program. Hank Rampton made an attempt to establish a foundation planting but by this time there was little need for this. Over the years the suggestion has been made that Oregon Ladino white clover is a tetraploid and the regular white clover is the diploid. I know of no proof to substantiate this.

As with red clover, there is proof that white clover does carry viruses which can affect the seed and forage yields. These may very well be seed transmitted. Very little research has been carried out on the viruses that would give a definitive answer.

CRIMSON CLOVER

Crimson clover was really not a factor in the 1930s as far as seed production was concerned. One account states: "Several years of effort on this have not been very productive. Farmers have several objections to the crop, the main one being that it takes two years to get one crop of seed."

Something must have happened as the Oregon acreage increased to 10,000 acres during the war years. Unlike red clover, Crimson clover is a winter annual to be planted in the Willamette Valley in September or October. One of the first clovers to bloom in the spring (May 15th), Crimson clover needs to be one of the first crops to be inspected.

In Oregon crimson clover was first used as a cover crop in orchards. However, in the south, Crimson clover was used as a perennial forage crop and a provider of nitrogen during the war years. As long as the seed planted had a high hard seed content, the crop would re-establish itself year after year in the south.

By 1957 there were over 10,000 acres of Crimson clover certified. The variety Dixie was the only variety. Dixie was a variety that was released from Georgia and was probably an ecotype. Very little breeding or selection work went into this variety. Loren Smith, a grower south of Corvallis, was successful in obtaining Breeders seed so he grew foundation seed for many years. It was enough to supply the Willamette Valley during that time when we certified over 10,000 acres.

For political reasons the Agriculture Stabilization and Conservation Service offices in the South started to make noises that they were going to stop paying growers for seeding this variety, if the seed came from Oregon. Dixie did not contain enough hard seed, they claimed. There really was no truth to this, but that was the only excuse ever heard. Eventually, the acreage dropped drastically until it was found that Crimson clover was in demand overseas, namely in Italy. A couple of new varieties from Mississippi came upon the scene, but the acreage was small.

Crimson clover was an easy crop to grow and certify. Most of the acreage was in Linn and Yamhill counties. Isolation was seldom a problem. During some cool wet springs, Southern anthracnose could be found. To my knowledge it was never proven that it was seed borne so no stigma arose out of our finding the disease in our fields. Normally we began our field inspections on the 15th of May in Yamhill county. It was a hard way to begin the season, but with one of the county agents driving and two of us never seated for long, we toughened up in a hurry. Rain made it more difficult of course. But, by June 1st we were usually through with Crimson clover inspections.

Weeds were a problem in Crimson clover. Cranesbill was listed as noxious in the southern states, so for years Cranesbill was listed as a no-no in certified Crimson clover seed. Unfortunately, considerable Crimson clover was not tagged because it contained cutleaf Cranesbill. The seed trade informed us that the cutleaf was not the one the southern states worried about, it was Carolina Cranesbill. Attention then shifted to yellow mustard or rape. Suddenly our field inspections took on new meaning. There is no way of proving this, but it is my contention that the lack of hard seed was not the reason Oregon seed was discriminated against, but the presence of mustard seed and this mainly in the uncertified lots.

Crimson clover was tried one year in Central Oregon, but with limited success. No doubt the winters would be too severe.

TREFOILS

Perhaps not too much can be said about trefoil production in Oregon. The crop is not too difficult to grow, but does shatter as the plant approaches maturity. Many unique ideas have been tried for harvesting trefoil. Don Hector used to have a large field of Cascade Birdsfoot Trefoil in the bottom across the highway from the Hylsop Experimental Farm. Don's procedure was to cut and windrow the crop early before ripe and would come along a few days later and loosely bale. The seed was allowed to cure and ripen in the bale, and then the bales were fed into a stationary combine. The Soil Conservation Service and perhaps others windrowed the crop onto paper strips, allowed the crop to ripen, and then combined taking in the paper and all.

In 1946, Extension Bulletin #694 shows 1,200 acres of lotus harvested for seed. Of these, 775 were in Jackson county and were probably Lotus corniculatus or Birdsfoot Trefoil. There were 250 acres in Clatsop Experiment Station. There were two types of big trefoil: hairy and smooth. Beaver was a hairy type and Columbia was a smooth type.

The Soil Conservation Service was responsible for selecting and releasing several trefoil varieties over the years.

ALFALFA

Alfalfa was grown for seed in Oregon prior to 1900, although not a popular seed until 1937. The first forage seed to be certified in Oregon in 1924 was Grimm Alfalfa. Experimental work on alfalfa for seed was being carried out as far back as 1912. The Metolius Experiment Station reported "Part of the field has been clipped off and part has been allowed to set for seed and in numerous instances this first year alfalfa has set a considerable amount of seed."

During the years of 1929-32, Mr. E.R. Jackman helped the alfalfa growers of Oregon find registered and blue tag seed of Grimm and Ladak Alfalfa. He quoted prices for Blue, Red, and Yellow tag seed and Growers Associations in Montana and Idaho where seed could be obtained. By 1936, there was 555 acres of Ladak. Most of this acreage was in Malheur, Baker, and Union counties. The 1939 report mentions two other varieties of alfalfa, Orestan and Cossack. Mr. Jackman, reporting on the Alfalfa Society of Agronomy meetings November 22-24, 1939, states that: "When bacterial wilt became widespread and work at a dozen stations indicated that the only hope of combating it lay in resistant varieties, the U.S.D.A. began to search for foreign varieties that had been growing in wilt-infested areas for generations. The result was the introduction of hundreds of strains of Turkestan, Persian, and Indian alfalfas. Most of these, of course, proved unadapted here, but some were satisfactory over wide areas. The first of these to be developed commercially was Orestan, a variety chosen by Mr. Larson, county agent at Ontario. It is now grown on about 1,000 acres in Eastern Oregon."

By 1944, production had fallen to 290,000 pounds. A tremendous leap in production in five years, occurred from 1948-52, when a record of 3,675,000 pounds were produced in 1952. Not all of this was certified, of course. Oregon has not been important among the states as a producer of alfalfa seed. It ranked number 12 nationally in 1958. While the acreage has not been great, the seed growers have been able to obtain excellent seed yields. This was made possible through the use of pollinators. Honey bees are not very efficient pollinators of alfalfa, but thanks to some observant growers, and Dr. W.P. Stephen, an OSU entomologist, some other type of bees were found that were much more efficient. The Alkali and Leaf Cutter bees were introduced for pollination and seed yields increased dramatically.

In the late '50s and '60s Oregon was contacted to see if they would grow Foundation Vernal for the National Foundation Seed Project. Several acres were planted. The first harvest showed the presence of red clover and a subsequent harvest ended with an immature seed of dodder showing on the seed test.

Overnight, dodder became a serious problem. Prior to 1959, alfalfa fields were allowed to have plants of dodder growing in them. In 1957, Ed Hardin started to work at the Oregon State University Seed Laboratory coming from Washington State. Ed acquired a small velvet roll and began running a full pound of alfalfa seed across this mill. The result was that a considerable number of seed lots that would have previously passed the requirements were now being marked no-tag because of the dodder seed. The alfalfa seed cleaners knew how close they needed to clean in order to get a lot passed, so when a full pound was examined, that changed the complexion considerably.

This meant that field inspections were going to have to be tightened. The growers were warned for a year and then we instigated the new measures: if dodder was found in the field, the field was rejected, however, the grower could go out and cut the dodder plants and burn them asking for a subsequent reinspection. Fortunately, it was discovered that Chloro IPC sprayed on the fields after the first cutting of hay was taken suppressed the dodder. These first few years of field inspection were not fun. We had very little support. Perhaps Mr. Ted Avery or Mr. Dierking were about all we could count on. The growers kept asking why the Idaho growers were not asked to do all of this. The Idaho Crop Improvement Association was not running a full pound of seed and they were not requiring dodder to be removed from the fields.

Rex Warren, Seed Extension Specialist, checked with his counterparts in some consuming states to discover their attitude about a noxious weed (dodder) being in certified alfalfa. Basically everyone contacted said they would recommend against using this seed. Also, at about this time, word was circulating that Utah had ignored dodder in their seed fields for years and now the dodder was so heavy that it was reducing yields and could not be totally cleaned out of the final product. The growers eventually accepted our inspection procedures even after we added Canada Thistle to the list of no-no's. The use of a heli-copter for making field inspections certainly made it easier to spot dod-der patches.

The certified alfalfa seed growers were a demanding group to work with. Many of the growers were growing certified seed in both Oregon and Idaho and Oregon and Washington. This meant that the growers could play one state against the other and this was in fact what took place. Because of the Oregon growers' attitude, New York declined to give them seed stock of many of the new New York varieties. Somehow this was construed to be caused by Oregon's Seed Certification rules, not because the alfalfa growers through the National Farmers Organization tried to lobby the New York legislature into changing Cornell's practices of Foundation Seed Release.

Although no written evidence has been found, Oregon State University has been given credit for establishing a procedure for Oregon seed growers, of selling their seed on a clean seed basis. Most other seed producers were selling on a basis of in-weight and a percent-age of clean-out (clean seed) as determined by the processor of the seed. This procedure eventually helped establish hundreds of private seed cleaning operations throughout western Oregon. This same poli-cy of selling on a cleaned seed basis could not be implemented with the alfalfa growers. The majority of the alfalfa seed was cleaned in Idaho or Washington. In these two states the grower would deliver his seed, a sample was drawn and the grower was paid on the basis of certified seed, only on the estimated total clean seed assuming the field was passed. A grower may decide later to sell his clean and tagged seed to another dealer. If this was the case, the company that processed the growers seed was to send the seed to the buying dealer. However, at least one grower found that his seed was not shipped, but supposedly comparable seed was shipped. With all this knowledge the growers came to the Seed Certification office in 1964 and asked that certified seed be sold only on a clean seed basis in Oregon. This meant that the Seed Certification office insisted on a purity and germination report on each lot of seed before tagging.

At this point, around mid-1960s, it was decided that it was time to open a Seed Certification office in Ontario, Oregon. George Tiger was hired after graduating in Crop Science and spending several summers making field inspections as a student. Although George was housed in the Malheur County agents office, his responsibilities covered all of central and eastern Oregon. This put George on the road a great deal and it was expensive, however; it did seem to satisfy the alfalfa growers. It did create problems for the Seed Laboratory as the Eastern Oregon growers started to lobby for a satellite lab in Ontario or LaGrande.

Through the efforts of Wally Rice, manager of Josephine Growers Association, there was a constant number of certified alfalfa acres in Josephine and Jackson County. The variety was called Talent and was a release from the Southern Oregon Experiment Station. It was probably the only variety grown in Oregon that was adapted to the southern region of the U.S., versus the Northern winter hardy varieties. Somehow Wally found a market for this variety overseas in Greece. The hoops that Wally had to go through to get the seed shipped and accepted by Greece were many but none-the-less it was very profitable for the southern Oregon growers.

It should also be mentioned that E.F. Burlingham & Sons at Forest Grove owned some land just west of Pendleton. Bill Cyrus, a former Washington county agent hired by Burlingham, discovered a few alfalfa plants with rhizomes on this land and named a release as "Nomad." This was one of the first private varieties of alfalfa released in the U.S.— one of hundreds that were to follow several years later.

PEA AND VETCH PRODUCTION

The following notes taken from a class presentation by Professor George Hyslop spells out how Oregon got involved in the Austrian winter field pea production.

AUSTRIAN WINTER FIELD PEAS (*Pisum arvense L.*)
—Information taken from Professor George Hyslop Class Notes:

"Austrian winter field peas are one of the most important seed crops in Oregon. Seed for this crop was introduced into the United States from central Europe for use in producing soil improvement and forage crops in the eastern and southeastern states. They are winter-hardy in the South and in the areas west of the Cascades, in Washington and

Oregon. Austrian winter peas are grown for the production of forage and seed to a limited extent in New York, Michigan, Wisconsin, Minnesota, South Dakota, eastern North Dakota, Montana, Colorado, Idaho, California, and Washington. Of this group of states, Idaho, Washington, and California, with a combined production of 9,000 to 11,000 acres, are by far the most important. However, commercial production of Austrian winter pea seed in the United States is relatively unimportant outside Oregon. In 1940, this state produced 80% to 90% of the domestic crop. They are grown as a cover crop and green manure in most of the cotton belt, and for the same purposes in certain sections of the Pacific Northwest. A limited amount of seed is produced in the South. However, seed production there is very uncertain and is generally considered uneconomic.

Austrian winter field peas have many desirable characteristics for soil improvement and forage purposes in the South. They make a very satisfactory growth during the winter and early spring, decompose readily when turned under, and are relatively high in nitrogen. Under favorable conditions green weight yields of 7,000 and 14,000 pounds per acre are obtained. As a forage crop they are high in nutritive value, very palatable and can be utilized as pasture, hay, or silage. Seed yields in Oregon from 1936 to 1940 averaged about 760 pounds per acre.

Austrian winter peas will not survive the winter in the colder sections of the United States. However, a cool growing season is required and high temperatures are especially disastrous when pods are setting.

Initial plantings in Oregon consisted of two 1/20th acre plots sown in 1923 at the Agricultural Experiment Station in Corvallis. In 1927, the crop assumed some commercial importance in the state when 250 acres were harvested. From that date, the acreage has increased very rapidly to the all-time high for the state of approximately 65,000 acres harvested for seed in 1940. The value of the 1940 seed crop in Oregon was about $1,400,000."

AUSTRIAN WINTER PEA SEED—E.R. Jackman Annual Report

"Beginning in 1929, the production of Austrian winter peas for shipments to southern states became a very active Extension project. The specialist located seed supplies, gave information upon production, and contacted prospective buyers in the southern states. Many hundreds of letters were written, newspaper articles prepared, radio talks given, etc. The acreage increased rapidly until 1931, when more than 11,000 acres were planted. At this time, heavy weevil damage began to appear so that farmers went to the work of growing a crop only to have the seed destroyed by weevils. The freeze in the winter of 1932 killed most of the peas, so the 1933 acreage was very light. The peas have had a tendency to migrate from year to year to weevil-free territory.

Work on garden peas in 1937 by the Oregon Experiment Station showed conclusively that rotenone dust, properly applied, would eliminate pea weevils in the garden peas, and it was reasonable to assume that this same procedure would control the weevils in the Austrian winter peas. Accordingly, extensive weevil control preparations were made by county agents in all counties growing peas. Meetings were called of all the growers. The method of control was thoroughly discussed and cooperators agreed to buy or build dusting machines properly equipped with wide hoods. An estimate was made of the number of acres in each community which each duster would serve, and if it became evident that there would not be enough dusters in any community, the matter was taken up with growers there, and some additional men would volunteer to put in this equipment. Dusting equipment was available for nearly every grower and at the harvest time the value of

Winter peas

the work was apparent to everyone. Seed cleaners could tell immediately when a lot of peas came in from a field which was not dusted. In some counties with several thousand acres, there would only be three or four small fields which were not dusted.

The large expansion of peas in 1937 and 1938 was very opportune in that the AAA program in the southern states demanded increased supplies of the winter pea seed. The entire project has been a good one, not only for the Willamette Valley, but for the southern states."

The records indicate that in 1926 in Benton County, there were 50 acres of peas grown for seed. By 1930 that acreage had jumped to 6,132 acres grown throughout western Oregon. 1934 saw a reduction to 4,025 and in 1937, the acreage really mushroomed to 26,326. The five-year average, 1940-44 indicates 55,500 acres grown for seed. Most of this acreage was still in western Oregon, however, the northeastern counties of Baker, Malheur, Union, and Wallowa grew nearly 10,000 acres.

By the mid to late '40s, the acreage had dropped to 20,000 and stayed at this level through the late '50s.

Harvesting field peas in the Mid-Willamette Valley

HAIRY, COMMON, AND WILLAMETTE VETCH
—Information taken from Professor George Hyslop Class Notes:

HAIRY VETCH

Hairy vetch (*V. villosa*), including the smooth stemmed variety, is a very popular cover and green manure crop in the South and is an important seed crop in Oregon. In the South, hairy vetch is sometimes known as winter vetch or Russian vetch. Hairy vetch is very winter hardy and withstands winter temperatures in the northern part of the United States except in ground that is otherwise bare and lacking protection in the winter. The hardness of the seed permits self-seeding and accounts for the volunteer plants on grassland along roadsides and grain fields.

Hairy vetch in the South is more resistant to stem rot than Austrian winter peas, but because of later development is more susceptible to attacks on the spring brood of the corn ear worm before it is ready to plow under. Small amounts of hairy vetch seed for local use are produced in the northern part of the cotton belt. However, seed yields are generally variable and undependable in the South.

Vetch and oats for silage

The commercial production of hairy vetch seed in America dates back to 1915. Prior to that time, a few growers in Michigan and scattered localities in other states saved some seed, mostly for local consumption. During the first World War when seed imports were discontinued from Europe the commercial production of hairy vetch seed assumed considerable proportions in the state of Michigan. During the years 1915-1919, about one million pounds were produced annually in that state. Small amounts of seed were also produced during that period in Indiana, Ohio, New York, Pennsylvania, Connecticut, Virginia, North Carolina, Georgia, Alabama, and Oregon. None of these states were self-supporting in vetch seed requirements nor were they considered important as producing areas.

While hairy vetch has been grown in limited quantities in many sections of the United States for more than thirty years and has many favorable characteristics, its production for seed has been practically discontinued in all of the states listed above except Michigan, and the 1940 production in that state was approximately one-half the 1915-19 average (500,000 pounds). In 1940, Washington and Arkansas produced 600,000 to 700,000 pounds of hairy vetch seed. For the past ten years, Oregon has produced 90 to 95% of the commercial seed crop in the United States. From 1936 to 1940, inclusive, seed yields in this state averaged 250 pounds per acres."

Harvested Hairy vetch

With the price of vetch being twice to three times that of peas, it is not surprising that vetch acreage also tripled. The four-year average between 1936-39 shows 33,600 acres of Hairy vetch and 9,800 acres of common or Willamette vetch. Between 1940-44, the average was 96,200 acres of Hairy vetch and 51,500 common or Willamette. However, the trend of species had reversed in 1945 as there were only 47,000 acres of Hairy vetch and 95,000 acres of Common or Willamette. The 1950s saw the acreages even out with 70,000 acres of Hairy and 77,000 acres of Willamette and Common."

Early history of the Farm Crops Department indicate testing on vetch as early as 1900.

The production of peas and vetch seed in the Willamette Valley was big during the Second World War years. This tapered off in the late 1950s. From all accounts, most of the early production was certified. Mr. Jackman addressed the Willamette Valley county agents: "The inspections began early in the year and progressed until spring break when all the members of the Farm Crops Department turned out to finish the job. There was only one field inspection conducted and the volunteers of different species between the rows were counted against the field. Isolation was not a problem. As more and more acreage was planted, it was inevitable that the hard seed would cause problems by the 1950s. Very little acreage was certified."

It appears that in order to qualify for the AAA War Production, the seed had to be certified, fumigated, and tagged. Probably not tagged with a certified blue tag, but a War Production tag of some kind. All of the major seed cleaning plants in the Willamette Valley built fumigation rooms to accommodate the 50 million pounds that were produced during the high-producing years.

Chapter V

Potatoes

From our earliest records, it appears that potatoes were the first crop grown in Oregon for food consumption. Quoting from the Dictionary of Oregon History, "the first recorded planting of potatoes in Oregon Country was made by the crew of the ship "Ruby" under Captain Bishop on an island in the Columbia River near Cape Disappointment in 1795. At Fort Astoria twelve shrivelled potatoes, all that remained of a supply brought from New York by the Astor ship, "Tonquin," were planted in May 1811. These produced 190 potatoes the first season and permitted the sending of a few plants to inland traders. In 1812, 50 to 60 hills planted at the fort produced 5 bushels. In 1913, 2 bushels planted produced 50 bushels. At Fort Vancouver, 1,300 bushels of potatoes were produced in 1835. From the time farming first began at Fort Astoria until enough wheat was raised to support the inhabitants, potatoes were the main substitute for bread.

An Indian chieftain, to whom a few of the tubers were given, failed to see any advantage in what they termed, Boston Root, over their own popular Wapato Root and did not go in for potato cultivation except in a desultory manner. With pioneer settlement, potatoes became a staple of diet. In the Willamette Valley in the 1880s, extensive cultivation brought fame to a Mr. John Dimick as "Potato King."

One of the first Oregon Agriculture Experiment Station Bulletins #30, authored by H.T. French, dated March, 1894, had this to report on potatoes: "One hundred and fifty-three varieties of potatoes were grown last season on the Experiment Station Farm. Eighty-eight were grown in plats large enough to warrant a calculation of yield per acre. The remainder were grown in small amounts to determine their manner of growth as affected by our soil and climate. The potatoes were grown on a clover sod which was plowed in the winter and again in the spring just before planting. No fertilizer was used except a dressing of wood ashes sowed on the field after planting. The potatoes were cultivated with a Planet Junior one-horse cultivator, thus keeping the ground as level as possible. While the season was unfavorable for the potato crop, owing to the cold dry weather, the yield was fairly good and the quality of the tubers was very good. There was very little fungus growth or dry rot on the potatoes last season. A thousand one-pound packages of potatoes were sent out to the farmers of the state last season for trial and in several cases the results were unsatisfactory because of the fact that fresh manure was plowed into the ground in spring just before planting. Manure applied in this way makes the best condition possible for the development of fungus diseases.

The Central Oregon Metolius Experiment Station reported in August of 1912, "That the potatoes were examined above ground a few were examined below ground. In the main, the potatoes look very well, but it seems that a little hilling would be beneficial in a number of cases, as some of the potatoes protrude above the surface. It is impossible to estimate the yield of the various potatoes, but some of them will be very productive if the hills examined are representative." What followed were some of the varieties that were being tested, as well as some comments:

Prize Taker – This was a late variety having a good vine.

Early Eureka – Not very promising looking, having a smooth white skin and rather irregular shape. *Earliest of All*, flat half round potato with white skin and rather shallow eyed. It is ripe now and has made a good growth.

Sir Walter Raleigh variety – has a dark green purple flower, erect plant and vigorous in appearance. It has a smooth, slightly netted flattened tuber which is a little longer than a round type.

Rural New Yorker – yellowish medium long to a half a round shape. Inclined to be little longer than the Sir Walter Raleigh variety. Pronounced netting of the skin, slight trace of pink in the eyes.

Early Sunrise – local seed. These are pink potatoes and in my opinion are some type of early Ohio, at least they are not Early Sunrise, which is a long white potato.

White Ohio – a white potato blotched with pink, having a slight netting. Pink eyes which are medium deep. The shape is somewhat like that of the early Ohio. It does not appear very vigorous. It is ripening now.

Early Ohio – a pink potato with rather deep eyes, somewhat cylindrical and not very smooth.

Gold Coin – late, but very vigorous. It is somewhat flat. It is of medium length, has medium number of white eyes, and a smooth white skin with slight netting.

Improved Rose – has a good vine growth, but is somewhat late.
A few other varieties were not tested as they do not appear to be very promising.

W.S. Carpenter, Extension Specialist Farm Crops, reporting in his annual report, November, 1919 to November 1, 1920, that he had distributed to the following counties twenty sacks of certified potatoes to Morrow County, 1,050 sacks to Malheur County, 15 sacks to Lincoln County, 9 sacks to Josephine County, for a total of 1,094 sacks. We're not told which varieties these were or the amount of pounds per sack. However, the same man reporting on the Seed Certification work on potatoes those same years, has this to say, "Seed potato and seed wheat certification has been a major feature of Farm Crops Extension activities. The object of Seed Certification is to develop a reliable source of the best seed obtainable in Oregon. Potato Certification - it is not possible at this time to summarize the results of this seasons certification of potatoes since the bin inspections have not been completed." The following summary of the first field inspection in the various counties shows the extent of the work. . . . "In the summary of the first field inspection of potatoes, the number of growers reached was 80. Acreage inspected, 180, acreage rejected, 72, for a total of 208 acres. The second and last field inspection has just recently been completed. Fifty-five acres of potatoes will be entered for the bin inspection. The whole value of potato certification work is not indicated by a statement of the acreage inspected or of the bushels certified. Eighty potato growers were given instruction by demonstration in their own field on the following matters:

1. Location and identification of diseases.
2. Location and identification of varietal mixture.
3. Roguing for disease and varietal mixture.
4. Hill selection methods.

Wherever possible, the inspection was made in the presence of as many potato growers as could be reached. The first field inspection in Multnomah County was conducted in the manner of a demonstration tour. Twelve potato club boys and eight potato growers accompanied the inspector and county agent and were given all the assistance possible in so short a time."

Mr. Jackman, in his annual report as an Extension Specialist in Farm Crops from July, 1922 to December of 1922, makes this comment about potatoes; "Potato Seed improvement work is much more complicated than that of grain work due to the numerous potato diseases, which in one year's time may destroy a seed strain which it has taken years to build up. In only two counties are there large amounts of comparatively disease free potatoes, these are Umatilla and Deschutes. Before this project can succeed there must be a large amount of educational work in convincing growers of the losses due to poor seed. In the field, of course, these two parts are so closely woven together that they are inseparable. In the following counties, part of the goal has been achieved in that large numbers of growers are awake to the losses from poor seed, but there is not enough really good seed supply to meet the demand: Malheur, Union, Wasco, Crook, Multnomah, Clackamas, Washington, Columbia, Josephine, and Benton counties.

There are eight counties growing potatoes commercially where growers are not sufficiently alive to the seed problem. Sixteen counties do not grow large amounts of potatoes commercially."

Mr. Jackman, in his 1924 annual report, makes a very good point. He is summarizing his 1924 season, "This year in potato certification a new grade of seed called, Standard Seed, was included for the first time. This is seed which is not as good as certified, but which is high enough quality to make very dependable stock for commercial growers. Formerly many growers became discouraged because of their inability to produce certified seed and quickly lost interest in the entire project. It was thought with the possibility of obtaining various degrees of success, the interest of growers would be better and they would stay in this game if they failed to produce certified seed after a year or two of trying. Indications are that the new grade has justified itself. Potato work was started for the first time in Klamath and Lake counties. It is believed that Klamath County especially will eventually be among the leading counties in the state in potato exports due to a favorable freight rate, soil and climate and an abundance of irrigation water. A great deal of interest followed the importation of a carload of certified Netted Gem seed in the spring of 1924 and the best potato tour ever attended by the specialist was held by county agent Henderson. The county has been thoroughly sold on certain irrigation and cultural practices as a result of this years work and next year it is planned to demonstrate more fully the value of good seed. A very successful certification year was concluded in Washington County where some high class Burbank seed exists. The project is fast making headway in that county. In Benton County, also, some progress has been made. In Multnomah, Clackamas, Columbia, and other valley counties, the situation is not satisfactory due partially to the lack of a central testing field for all strains of certified seed. The inspector believes that this is a necessity before real progress can be made with seed improvement in Western Oregon. Because of the limited field force, it is often impossible to visit fields when they should be visited and for this reason it is not always possible to see fields at the proper time for making inspections. A central field where all the seed would be tested would greatly overcome this handicap. In that case, visits to this field could be made frequently and any mistakes in passing upon fields would be detected at once. The certification in Umatilla County was eminently satisfactory. A large acreage passed as Standard Seed and the majority of growers are feeling very good about the results of their work in roguing, hill selection, and so forth during the past years. These producers of Standard Seed have already contracted to sell their seed at a minimum premium of $10.00 per ton over table stock. In former years the premium received was small or none at all. Quite a good program in seed improvement was started in Lake and Wasco counties, but largely nullified by the driest summer in the history of the state. It will be entirely impossible to bring potato work up to its point of greatest usefulness until at least one more specialist can be employed."

This same year Mr. Jackman put out a grower's list and a comment which is interesting, "The Portland Seed Company during the last three years has made a systematic and intelligent attempt to build up some reliable strains of potato seed. This is the only seed company that I know of in the Northwest that is doing this kind of work. This year for the first time they have a considerable amount of seed to sell. I want to call your attention particularly to the Earliest-Of-All seed, which is the most reliable of that variety. They are offering their certified Earliest-Of-All in 50 sack lots at $2.50 for immediate sale and subject to prior orders. In less than 50 sack lots they are charging $2.75. As this is written they have some small seed of that variety which they are offering in large lots of $25.00 a ton. This is probably the best seed potato buy in the state, as the seed is small because of the late planting and not through sorting it out of large seed. In counties where a large amount of this can be used, I can recommend this lot highly both because of its quality and because of its cheapness. It is much superior to ordinary seed. The above prices are all subject to prompt acceptance and to prior order. Those interested should call or write, L.W. Wheeler, Portland Seed Company."

Mr. George Hyslop, Extension Specialist Farm Crops in his 1928 annual reports makes these comments concerning the first Oregon State Potato Growers Association. "The state is becoming very well standardized upon Burbanks, Netted Gems, Earliest-Of-All, or Idaho Rurals as the main crop Potatoes. Some attention is being given to Irish Cobblers, Bliss Triumphs, and Blossom White Rose. The industry is gaining in importance. The outstanding thing in connection with potato improvement work in Oregon in 1928 was the organization of several additional potato growers associations and the Oregon State Potato Growers Association. The latter held its initial meeting on July 7 at Prineville and held another meeting at the Pacific International. It has sponsored affiliation with the National Potato Institute and was the sponsor in the state legislature for the new potato grading law. Association members have participated in conferences in Pendleton, Oregon; and Spokane, Wapato, and Walla Walla, Washington, in the interest of getting more uniform grading practices in the Pacific

Northwest. Potato Shows have been held in Umatilla, Deschutes, Klamath, Clackamas, Washington, and Columbia counties under the auspices of the various local associations. Various members participated also in the general shows at Pacific International and at the Northwest Potato Show in Spokane. Oregon won honors in certified and commercial Netted Gems and Burbanks at the Spokane Show. A sample of the first prized Netted Gem commercial potatoes grown by Henry Semon of Klamath Falls was forwarded by airmail to President Coolidge. Roy Roberts of Powell Butte won first on certified Netted Gems and Rowell Brothers of Scholls won first on Burbanks at both Spokane and Portland." It is interesting to note some of the happenings as reported in the 1932 annual report for Mr. E.R. Jackman, "Some of the results of work to date on potato projects, 1) Eastern Oregon Standardize 100% on Netted Gems as a late crop commercial variety. 2) Klamath County developed into one of the leading Netted Gem producing areas. It changed in seven years from a strikingly home consumption area to a region selling on the average over one million dollars worth outside of the state. 3) There are grower's education associations in Baker, Umatilla, Klamath, Crook, Clackamas, Columbia, and Washington counties. The specialist is the secretary of a state-wide growers educational association. 4) Every commercial grower treats with a corrosive sublimate or hot formaldehyde except a few new growers in Klamath. 5) A start made in developing a steady seed trade with California. 6) Oregon potato production is coming back and car lot shipments are increasing. 7) Western Oregon completely standardized on Burbanks for a late crop."

By 1939 some trends had started to take place in the potato production in the state of Oregon. The program has attempted each year to obtain sufficient acreage of potatoes to supply good quality seed for local use and to increase the varieties certified as markets developed. Assistance is also given to the marketing of potatoes in Crook, Deschutes, Klamath, and Multnomah counties. This past year ring rot disease losses obtained major proportions in Malheur county. Assistance was given growers in this county in an effort to sort their stock to eliminate all possible losses from breakdown in transit. Growers were assisted in roguing seed plots in an effort to obtain as clean of seed as possible for the next year's planting. The principal increase in acreage of certified potatoes was in Multnomah County. Dr. M.B. McKay, formerly U.S. Department of Agriculture plant pathologist on potato disease, has developed very excellent stock of White Rose seed potatoes. He is rapidly obtaining a wide reputation for his seed, practically all of the White Rose potatoes in this section were marketed through Dr. McKay. The past year he grew about 40 acres of his potatoes in tuber units (four or five plants from the same seed tuber in a row) and rogued carefully. From this planting, he supplied most of the planting stock for all of the farmers in the area. Dr. McKay plans on growing all of his potatoes in the tuber units this next year and is requiring that all farmers who market through his organization grow theirs in tuber units also. With increasing threat from ring rot, the potato certification procedures will be changed next year to include additional field inspections and to require that each lot of potatoes before it is eligible for tagging have a laboratory culture to more definitely determine the presence or absence of ring rot. This additional work will necessitate some increase in potato certification inspection fees. In 1939 Professor Hyslop made the field and bin inspections in Klamath and Lake counties and the assistant crop specialist took care of this work in the Willamette Valley. LeRoy Hansen and Harold Finnell of the Farm Crops Department assisted with some of the work in Klamath, Lake, and Willamette Valley counties.

In 1939 before potatoes were eligible for tagging, they had to have a laboratory culture run on them to ascertain the presence or absence of ring rot disease. At some point shortly after this, perhaps during the war years, many samples of seed were shipped and planted on the Camp Pendleton Marine Base at Oceanside, California. This location allowed for mid-winter testing for leaf roll and mosaic. The procedure was to take a sample off of the pile of certified potatoes in storage. These would be held until dormancy had passed and then planted in rows at Oceanside. The readings were normally made during the month of March or early April. Some of the drawbacks of this procedure were; occasional frost which set the whole disease reading schedule back. Fog and rain during the time when the readings were made caused the plants to mask disease symptoms and the readings were too late to benefit growers trying to get their seed crop to Northern California by January or February.

In 1947-48, the growers approached Oregon State College about the possibility of building six greenhouses to take the place of the Oceanside plantings. The potato committee of the Oregon Seed League in conjunction with the Oregon Potato Commission, were instrumental in getting the ball rolling. At the commission meeting held July 18, 1949, a motion was made to place $22,500 in the budget to provide matching monies for the $22,500 greenhouse appropriation by the State Legislature. Later on the commission agreed that if they were going to put up this money then any research work on potatoes would have first call on these greenhouses after the certification testing was completed. The construction bids came back high and the college decided that supplemental heating system would be needed, so only four greenhouses were built in 1950-51. Elmer Johnson, a new graduate of Oregon State College, planted the first crop in 1951-52. Several bugs had to be worked out i.e., breaking dormancy, cyanide gas was used first; size of tuber to plant, a melon baller was used for this; a planting board with melon ball size holes on two inch centers was made; and a temperature, moisture regime had to be worked out. Elmer Johnson must be given credit for working out the basics.

Depending when the samples arrived at the greenhouse, the readings could usually be made 60 days after planting. The disease symptoms were quite often obvious and the plants could be left growing for the growers to come and look at for themselves. Or at times they were present when the readings were made.

In addition to the winter grow outs, there was some eye indexing done in the greenhouse. This was not a complicated procedure, but seemed to be very beneficial for the growers. Usually these were the last to be planted so the dormancy did not have to be broken. An eye from the blossom end would be planted and an aluminum tab with a number on it would be inserted in the planted end and the same number in the remaining tuber. If a diseased plant was found, that numbered tuber was thrown out. Mr. Elmer Johnson left the program August 1, 1955. George Clark ran the potato program from July, 1955, to October, 1956. Jack Waud succeeded George Clark at that time. Jack worked full time for Seed Certification and completed a Master's Degree. In 1962, Jack left to continue his education, and it looked for awhile as though Don Brewer would take over, however;

in 1963, Bill Sieveking was asked to assume the responsibility of the potato certification program.

The State Seed Law, when rewritten, left the responsibility of Seed Certification with Oregon State College. However, the enforcement of the State Seed Law remained with the State Department of Agriculture. The enforcement of the Seed Certification rules was left up to the Dean of the School of Agriculture. Clearly an overlap existed. There really was no conflict except with potatoes. So, in 1945, a Memorandum of Agreement was instigated. Basically, the memorandum agreed that the State Department would do the Shipping Point Inspections, which they had responsibility for, and Oregon State College would do the field inspections and the greenhouse work. It could probably be said that the effectiveness of the memorandum was based on who was the Director of the State Department of Agriculture and the Dean of the School of Agriculture. Their attitudes dictated the effectiveness of the memorandum.

During the spring of 1966, Canada thistle was found growing in the peat soil on the dikes in Lower Klamath Lake. The Extension office in Klamath Falls recommended the use of Tordon. The weeds were sprayed and were killed. However, the residue from the spray was left on the peat dust and so every time a vehicle went down the ditch bank the dust kicked up and spread across the certified potato fields. By late August, the potatoes were showing very decided symptoms of Tordon damage. This was bad enough, but there was evidence to show that the Tordon would carry on through to the new tubers. So there was a big problem.

The Dean called a meeting with the affected growers, the Extension administration and Seed Certification. The conclusion reached was that Seed Certification was to systematically sample each affected field. These samples were to be grown in the greenhouses and any Tordon distortion in the potato plants would reject the fields for certification. The seed growers were certainly not pleased. Some of the Extension administrators felt we in Certification should have protected the county agents involved by ignoring the situation. The certification rules dictated that chemical damage had to be scored against the field. The sampling and growing of the tubers was a great deal of work and expense. Bill Sieveking's background was ideally suited for the task and is to be given credit for carrying this to the end in a very professional manner.

Bill left the potato program in 1968 and Oscar Gutbrod was asked to become the person in charge. The mid '60s and early '70s were disappointing years. Acreages were expanding with new and inexperienced growers, ring rot was being found after the fields were dug, Washington State was planting Oregon seed in grow out trials and reporting high incidents of leaf roll and mosaic. The validity of the greenhouse readings was being questioned. The acreage fees were being raised and, of course, this is never popular.

To Oscar Gutbrod's credit some new practices were instigated. By this time an advisory committee had been formed to advise Seed Certification. The commission with a great deal of success picked up one of Bill Sieveking's ideas of a statewide educational growers meeting and was working with Rex Warren, OSC Extension Seed Specialist on this.

Ring Rot trials were instigated, new and more restrictive standards were put in place. Better education of the seed growers was begun. Some of the program's biggest critics could not stay in the program and maintain the high standards called for.

Unlike the grasses or legumes, most of the Oregon-grown certified potatoes were planted in Oregon or adjoining states. This meant that any advisory committee for potatoes would have representation from both the producing and using segment. And this put the potato program coordinator right in the middle. The make-up of the advisory committee left Oscar as the secretary of the committee and the chairperson was always elected from one of the producing areas.

The costs associated with the production of an acre of certified potatoes was high, not only for the grower but the certification office as well. The result was that the grower had a great deal riding on the outcome of our findings and we had to be very positive of our readings. Again to Oscar's credit, he made sure that those who were working with the potato program were well trained. Thus, Oregon was one of the few states who stayed away from any major lawsuits and acquired a reputation of a well-run, aggressive certification program. Mr. Gutbrod continued in this program until his retirement in 2000.

Virus Testing Program

During the April, 1971 meeting of the Potato Advisory Council several presentations were made concerning a Leafroll Virus Testing Program. Dr. Youngberg outlined the basic goals of such a program, Dr. Koepsell presented the budget covering the initial costs and Oscar Gutbrod mentioned some of the seed certification standard changes that would become necessary if the program were to be instigated. It was agreed at this meeting that a leafroll free certification program should be implemented. The council agreed to an increase of $1.50 per acre to finance the initial operation on the 1971 certified acreage.

The council was told of the progress being made on the Latent leafroll virus-free potato seed stock at their February 25, 1972 meeting. At this point 450 plants had been established of the Russet Burbank variety with another 50 soon to be added. Progress was rapid enough the following year to warrant the asking for $15,000 to $20,000 to cover the cost of an isolation greenhouse. The greenhouse was to be built on the Hyslop Agronomy farm.

By January 1, 1976 progress was so rapid that Dr. Youngberg informed University administration that the Foundation Seed Project was to take over the responsibility of growing the Elite I and II classes of Virus tested material. Cold storage for the over-winter storage of the tubers was sought and obtained in the Withycombe complex on campus.

Arrangements were made by the Foundation Seed Project to use two acres on the Keith Cyrus Farm to grow the Elite I and II generations. Obviously this two-acre field had to be as isolated as possible.

By 1978 it was anticipated that Dr. Allen would be given 11 varieties of potatoes to heat treat and John Kelley would do the tissue culture work on these eleven. It was becoming obvious to all those working in the program that a full-time person was going to have to be hired to carry out and oversee the virus test, stem cutting program.

The Potato Commission did put up money at the request of the potato seed growers for the isolation greenhouse. People were hired to concentrate on reproducing meristem tissue that was virus free. All of this work was the responsibility of the Crop Science Department with cooperation from the Plant Pathology Department. Potato production as well as mint production were considered Farm Crops and as such were dealt with by the Crop Science Department. However, the work could not be conducted without the input from the Plant Pathology individuals such as Dr. Tom Allen and Dr. Paul Koepsell. Dr. Allen was doing research on viruses and thus was able to heat treat the tubers ridding them of the viruses. Dr. Koepsell was the Extension Plant Pathologist Specialist and advised the Crops Specialists on promoting the use and maintenance of these virus-free potatoes. Dr. Koepsell did have a very good working relationship with other potato researchers around the world. Using some very elaborate tools Dr. Koepsell found what he felt were indications of ring rot in the virus free material. Paul formally made this known September, 1985 after the crop had been grown and was ready for harvest. Needless to say this set the program back. Several growers went ahead and planted the tubers anyway; however, with the taint of ring rot the University could not release the material as Elite I and II. The program definitely suffered a set back and had to start over. None of the tubers planted from this suspected material was ever found to contain ring rot.

The Seed Certification Program for potatoes has had its detractors over the years. However, it certainly can be said that a considerable amount of goodwill for the University has been gained because of the program. There are a number of potato growers who would have been unaware of the Extension Service had it not been for the Potato Seed Certification emphasis. At one point in the 1970s the Dean undertook a reorganization of the certification program. This took the form of removing the Seed Certification work from the campus. This was so repugnant to the growers that they changed the Dean's mind very quickly. The potato growers were the most vocal over their distaste of this idea. It is reported that the Mayor of Redmond even called the Dean to register his dislike of the idea. The Dean decided this was not going to work and abandoned the concept.

Chapter VI

Grain Certification

Quoting from Oregon's First Century of Farming:

"Cereal grains were important pioneer crops with wheat being used largely for flour. The flour which can be spared from home use was sold to trappers and to the Hudson's Bay Company, thus becoming the first processed crop for marketing from Oregon farms. Oats and other feed grains gained importance as livestock numbers increased.

"Value of production of grain crops accounted for 43% of the value of all crops in 1899. In that year wheat was the most important food grain crop, representing 29% of the value of all crops. Oats were the most important feed crop while barley and corn were of minor importance. By 1958 the value of grain crops had dropped to 35% of the value of all crops. Wheat had dropped to 35% of the value of all crops. Wheat, accounting for a little more than a fifth of the value of all crops in 1958.

"Wheat: Wheat was one of the first and most important crops grown by the pioneer farmers of Oregon. The first planting was made in 1828 on the farm of Etinne Lucier. Following Lucier's successful planting of wheat, other fur trappers and employees of the fur company established plantings of wheat in the Willamette Valley. Prior to that time, potatoes had been Oregon's most important crop. They were used as a flour substitute. Wheat soon became a basic pioneer crop, being used for flour and seed during these early years by nearly every resident of the Willamette Valley."

In 1839 Oregon-grown wheat sold for around 60 cents per bushel. During the early years, the Hudson's Bay Company provided the only market for Oregon wheat. Because of the shortage of legal tender, they often accepted wheat in payment of promissory notes for seed and other supplies. The United States government, realizing that the company might require legal tender for payment of notes, made wheat legal tender with a standard value of $1.00 per bushel. Andrew Kilgore planted the first (commercial) wheat in eastern Oregon in 1863 on his Umatilla County farm, marking the first Oregon wheat produced outside the Willamette Valley. By 1909 the Columbia basin area exceeded, western Oregon's production. During the late 1870s and early 1880s, eastern Oregon's principal disadvantage was the distance from market with the resulting high transportation costs. Wheat shipped to Astoria was handled as often as ten times. The coming of the railroad helped eastern Oregon's position in the production and marketing of wheat. For example, this wheat previously had to travel 15,000 miles to reach the European market, being transported further than any other wheat in the world. The railroad made it possible to send wheat to the east coast before shipment to European markets. The opening of the Panama Canal in August, 1914 shortened the water route to markets in Europe and east coast ports, improving the international trade position of Oregon wheat.

Wheat production in Oregon continued to increase in the late 1800s with the upward trend reaching a peak in 1901 of nearly 16$1/2$ million bushels. This was followed by a five-year period of decreased production, with the 1906 harvest being only 10$1/2$ million bushels. Wheat production increased in the nine years from 1907 to 1915. During this period Oregon exported a large quantity of wheat. By 1912, Portland was exporting more wheat than any other port in the United States. Wheat production has fluctuated quite widely since 1915. This was due to weather conditions and to economic factors introduced by world conflicts. The general trend, however, has been upward.

The average yield of 36.0 bushels per acre in 1957 is the highest of record. Harvested acreage decreased under the farm program following the record year of 1953, until 1958 when an upturn again occurred. Production has increased since 1955 due mainly to higher yields, but still below the record production of over 34 million bushels in 1954.

Oregon's rank among the states in wheat production and acreage has usually varied between 14th and 20th during the past 100 years. Oregon ranked 16th in production and harvested acreage in 1958. Oregon ranks high as a producer of soft wheat and generally has a high rank in yield per acre for her total wheat crop. During the record production years of 1952 and 1953, Oregon ranked third and fifth, respectively, in yield per acre, while ranking 14th in production. In 1954, Umatilla County, Oregon, ranked eighth in production among the leading wheat producing counties of the Nation."

Dr. Hill in his recollections of "The Beginnings of Seed Certification in Oregon," believed the first certification in Oregon was with wheat. When the U.S. Grain Standards Act was passed in 1916, the wheat section of Oregon was producing several classes of wheat. There was a limited amount of Hard Red Spring, considerable Hard Red Winter, substantial amounts of Soft Red Winter, and three subclasses of white wheat, Hard winter, Soft winter, and White Club. The inspection records in 1918 showed that one third of all wheat inspected in Oregon graded as Mixed wheat, which, of course, had a reduced price. George Hyslop considered this to be an intolerable situation. With the help of some of the county agents from the wheat counties, a certification program was established. Basically, this was a program to keep the classes of wheat separate. Seed was obtained from farmers who were known to have clean seed. Field inspections were concerned mainly with mixtures of other classes of wheat, and with the presence of rye and certain noxious weeds, such as bindweed. By 1921, the inspection reports showed less than six percent mixture.

It should be emphasized that at this point there was little attempt to establish varietal purity. Later, as new varieties arrived on the scene, varietal purity became an integral part of the inspection procedure. It must be admitted that for a number of years, the release of new varieties was pretty haphazard. The superintendent of one of the stations, principally Moro or Pendleton, would let an interested grower have a quantity of seed. Sometimes this would be entered for certification, but more often not. Occasionally, the seed released was not up to certification standards, or it would be planted on fields which contained some volunteer wheat of other varieties. Dr. Hill recalls one summer in the early 1950s, all of the Elgin wheat entered for certification in Gilliam and Morrow counties was rejected. The seed had been obtained from the Union station. It was assumed that material from the branch station would meet requirements, but was not necessarily so. In this instance, I was the field inspector.

It should be borne in mind that in those days there was no certification department. The field inspections were made by the extension specialist and members of the Farm Crops Department. Although this was an extra activity, I found the field experience to be a most valuable one for my classroom activities.

When the variety Omar was released (I believe early in the 1950s), a new procedure was established. The three northwest states had agreed that new varieties adapted to the three states would be released simultaneously. Washington State University had provided Oregon with 400 bushels of Foundation seed of Omar. County agents were advised to select their better growers, that no allocation would exceed 10 bushels, and that the planting rate would not exceed 30 pounds. The largest allocation was made to Umatilla county. There were two reasons for this: growing conditions were better there and it was thought we might get a more rapid increase. The other reason was that the county agent, Vic Johnson, had done the best job with his growers. He conducted roguing schools and had alerted his farmers about the problems of certification. This created some backlash, as might be expected, especially from counties that did not get all of the seed requested. Umatilla county growers were especially vociferous about the 30 pound planting rate. The yield results from the 30 pound rate that year made believers out of many farmers."

Dr. Hill was being very candid with his remarks and reveals several points. First the tri-state seed release policy: When first instigated this policy worked well. Washington State had seen a need for a Foundation Seed organization and was able to get one started. It was not until the release of Gaines, the first dwarf variety of wheat, that the Foundation Seed Project (Washington Crop Improvement Association) actually started to make money and become a major player. Idaho handled their Foundation seed on an Experiment station and subsidized the production enabling their growers to obtain cheap seed. Seldom did they have much for Oregon and Washington growers. As Dr. Hill pointed out, any new releases that came from Oregon or to Oregon generally went through the county agents and often was placed on land that did not qualify for the production of certified seed.

In the 1960s Oregon State University released a six-row barley to be grown in the Willamette Valley taking the place of a two-row malting barley. The breeders seed was cleaned on the Hyslop Experiment Station where certain growers acquired their seed. The following summer at field inspection time Dr. Wilson Foote, OSU Cereal Breeder, asked to go along on the inspection trips. It soon became obvious as to why he was so interested. It turned out that sacks of two-row barley were mixed with the new six-row variety when the growers picked up their seed. Needless to say, this was a real source of embarrassment. However, it did point up a real need for an Oregon Foundation Seed Project. Thus, one was begun and housed with the Seed Certification program. It really never did have full support of the state's Experiment stations nor from Dr. Ritchie Cowan, Department Head of Farm Crops. After Wilson Foote's experience he was very supportive. The directors of the Malheur and Klamath Falls Experiment Stations had over the years obtained considerable local support and felt obligated to release any new varieties from their stations to the local growers. So, they were very cool to the idea of a Foundation Seed Project.

The Foundation Seed Project received a big boost when a California bean seed grower offered to give the program two complete lines of seed cleaning equipment just for the dismantling and hauling. The two air-screen machines were brought to Corvallis and stored. A young man who had just recently come to Corvallis with a good mechanical aptitude was hired to go through and refurbish the two machines. Eventually, the University gave the go ahead to build a building on the Hyslop farm for the housing of the seed cleaning equipment and seed storage. The building was paid for out of Seed Certification funds; with the expectation that the program would be able to pay the money back. Unfortunately, the project never did consistently make enough money to pay this sum back.

The reason for mentioning the Foundation Seed program under the Cereals heading is because it was anticipated that cereals would make up the majority of the volume that would pass through the Project. Indeed, this turned out to be true. However, it seemed that with every release of Foundation cereal seed there were major problems.

One of the major problems was a lawsuit that involved Washington cereal growers as the plaintiffs and Washington and Oregon Foundation Seed organizations as the defendants. Of course, this involved Oregon State University and the State of Oregon. The suit went against Oregon and Washington. The judge who summarized the jury's findings was of the opinion that the attorneys were really the only ones who gained by the suit.

After the smoke cleared it would appear that the problem all along was the fact that dwarf varieties of wheat were not totally self fertile, that there could be up to 12% out-crossing at the head row (pre-breeders) seed level; Actually, at all levels, if the fields were not kept isolated. There was insufficient space at any level to completely isolate all of the promising lines and so there were always problems.

In the early '70s when several new wheat varieties were being released, the eastern Oregon growers complained because they had no land that would qualify for growing foundation or registered seed. Under their cropping system, a year of summer fallow was insufficient to rid the soil of seeds of other varieties. The problem was more acute with the red kernel varieties then the white, even though the contrast was obvious in the field. Eventually, a system of seedling inspections was devised that seemed to take care of the problem.

As was indicated in the introduction, there are several thousands of acres of cereals grown in Oregon each year. Except for those beginning years that Dr. Hill mentions, very little cereal seed was certified in comparison to that which was planted yearly. The feeling amongst the growers was that they could select their own seed from their own fields or the neighbors. If changing varieties, they would go to their local grain warehouse and buy some seed. However, there was plenty of evidence to prove that smut and weeds were introduced to many a farm when seed was bought in this manner, but the growers just could not be convinced that certified seed was worth the extra money. Washington State seemed to have a little better success in selling their certification program. Being a Crop Improvement Association, they were allowed to advertise extensively. In Oregon it was difficult to get the Extension Service convinced that there was an advantage to planting certified seed. Even after a drill-box survey proved the point very well.

One of the strongest arguments the growers had against certified cereal seed was that certified seed coming in from Idaho almost always contained wild oats. This was an unthinkable weed in Oregon certified grain seed, but tolerated in Idaho seed. Since Idaho grew considerable spring grain anytime a need arose, the grain trade would go to Idaho for their needs and inevitably several truckloads would be rejected every spring. This situation pointed up the inconsistency between the three states. Considerable effort on Ron Cook's* part went into trying to standardize the three state programs on cereals. This was a task that sounds simpler than it turned out. With the two states, Washington and Idaho, having Crop Improvement Associations they could not make changes that were needed because their boards of directors might stand to lose some certified acreage and thus would not vote a change. This was exactly the situation with Idaho and wild oats. Many of the Directors on the Idaho Crop Improvement Association were from southern Idaho where wild oats grew profusely. They simply would have been out of the business if wild oats had been listed as objectionable. Instead, wild oats were listed as a separable crop which meant that they were not counted at the field inspection with the presumption they would be cleaned out before sale. With Idaho's system, the seed grain was tagged and shipped before the lab report was sent to the Crop Improvement office. Uniformity of standards in the three states were slowly developed, but with time and pressure from Oregon grain seed companies, the wild oat problem in certified seed has been reduced.

---

*After retirement of the author in 1989, Mr. Ron Cook assumed the head of the Oregon Seed Certification Extension Project.

Chapter VII

Forest Tree Seed Certification

The forest tree seed industry has been an active participant in international commerce since the beginning of the 20th century. Through the cooperative efforts of private tree seed interests and Scotland's Department of Forestry, seeds from similar ecological regions in Oregon and Washington were planted in trials in Scotland. As a result of these trials, commercial sales of tree seed were facilitated. Seed germination is difficult with conifer seed. In the early 1960s, John Cameron and Charlie Brown, private seed company representatives, began using the Oregon State University Seed Laboratory for viability testing. During that time, it was discovered that there was some real concern about where and how seed cones and hence seed were collected. On an average, there is only one good seed production year in seven. Given the foreign and domestic need for seed from certain areas, demand often outstripped supply. In fact, jokingly it was said that seed companies had shovels in the same bin all labeled with different areas and elevation according to the demand. Seed laboratory personnel, Louisa Jensen and Ed Hardin, suggested that perhaps a Seed Certification program for tree seeds could be of some help.

There was ample research to show that trees, in order to thrive, needed to be planted back in the same general climatic zone and elevation as that of their parentage. Consequently, when a forester replanted an area the zone and elevation were matched to that of the parentage.

In the 1970s, a Tree Seed Council was formed and Seed Certification standards were complete with zone maps, tags and field inspection procedures were adopted. It must be noted that the United States Forest Service Region 6 input was invaluable in this effort. The Forest Service and later the Bureau of Land Management (BLM) agreed that they would give preference to certified seed when awarding bids for procurement. Region 6 of the USFS encompassed both Oregon and Washington. Private seed companies collected cones in Oregon and processed in Washington or visa versa. Therefore, certification standards were identical for both states. In those early years, vast numbers of retired Forest Service personnel, who were dedicated to seeing this program carried out, were employed. Thus, finding individuals to make field inspections or audits was no problem. Mr. Lee Hunt coordinated the field inspection program for many years. Because of the international trade in Forest Tree Seed, Oregon and Washington's programs were watched with a great deal of interest. The United States was asked to join into the OECD program on the strength of these two programs. On several occasions, representatives of the two states were asked to attend the OECD meetings in Paris, France, to participate and describe their respective organizations. The Certification programs met the needs very well. Later the programs were expanded to include tracking the seed from storage to nursery plantings and eventually into the forest. This addition satisfied some of the environmental concerns.

Chapter VIII

Conclusion

The writer has attempted to summarize Oregon's early involvement in seed production. Much of the material was collected over several years. The summarization was written after my retirement while much was fresh in my mind. Great effort was given to credit those I have quoted. Much was gleaned from early Extension annual reports and conversations with Mr. Harry Schoth, Dr. D.D. Hill, Mr. E.R. Jackman and Mr. Rex Warren. Of course, there could have been more written about the OECD, and the Association of Official Seed Certifying Agencies (AOSCA), but those accountings can be found elsewhere. The underlying premise throughout is that the Seed Industry in Oregon would not be what it is today without the knowledge and hard work of the personnel of United States Department of Agriculture, Oregon State University Research, and Extension Services.

About the Author

This recounting of Oregon seed industry history flows from the candid pen of a true insider. Don Brewer's career with the Oregon Seed Certification Service placed him in a special position to view the evolution of that industry. His 32 year employment in this Oregon State University Extension Service program was interrupted only twice, once for United States military service in Korea, and the second time for advanced degree studies at the University of Missouri. Of those years, 25 were served as the Program Director of Oregon's Seed Certification Service. These were years of tremendous growth in the Certification Program, in crop types and varieties being certified, numbers of seed growers and companies participating, acres signed up for inspection, and pounds of seed tagged for both domestic and international markets. Don Brewer had direct involvement in and leadership of many new developments in the certified seed industry. His participation and leadership was here in Oregon, and also at national and international levels. He "came up through the ranks" of the Association of Official Seed Certifying Agencies (AOSCA), serving that organization as chair of many crop and organizational committees, and eventually as Director for sixteen years, Executive Director for ten years, and the AOSCA President. AOSCA recognized and honored Don by awarding him Honorary Membership in 1992. But this book of history is not about seed certification, nor AOSCA or the international Organization for Economic Cooperation and Development (OECD). Instead, it is a very personalized review of Oregon's emerging seed production industry; a ride along the early roads of an agricultural endeavor, meeting pioneering seed growers and seed marketers, and the early federal and state researchers and extension agents who helped the fledgling industry tackle new crops, overcome production problems, and develop new markets both here and abroad. Don's appreciation for people, love of history, and insight of behind-the-scenes-politics and a few seedy shenanigans, all combine to provide an enjoyable perspective. There are still enough "old timers" around who will recognize themselves, their ancestors, or a few unnamed others in Don Brewer's "Early History and Recollections of Oregon Seed Production."